Springer Series in Optical Sciences   Volume 18

Edited by David L. MacAdam

Springer Series in Optical Sciences

Edited by David L. MacAdam

Editorial Board: J. M. Enoch   D. L. MacAdam   A. L. Schawlow   T. Tamir

Holography in Medicine and Biology

Proceedings of the International Workshop,
Münster, Fed. Rep. of Germany, March 14–15, 1979

Editor
G. von Bally

With 224 Figures

Springer-Verlag Berlin Heidelberg GmbH 1979

Dipl.Phys. GERT VON BALLY
Experimentelle Audiologie,
Hals-Nasen-Ohrenklinik der Westfälischen Wilhelms-Universität,
D-4400 Münster, Fed. Rep. of Germany

ISBN 978-3-662-15811-1   ISBN 978-3-540-38961-3 (eBook)
DOI 10.1007/978-3-540-38961-3

Offset printing: Beltz Offsetdruck, Hemsbach/Bergstr. Bookbinding: J. Schäffer oHG, Grünstadt.
2153/3130-543210

Preface

The International Workshop on Holography in Medicine and Biology was held in Münster, Federal Republic of Germany, on March 14th and 15th, 1979, at the Clinic of Otorhinolaryngology of the Westfälische Wilhelms-Universität within the frame of the Symposium 79 of the Sonderforschungsbereich 88 "Teratology and Rehabilitation of Patients with Multiple Handicaps" of the Deutsche Forschungsgemeinschaft.

In fact, this workshop was not the first meeting dealing exclusively with biomedical applications of holography and related techniques. The very first symposium in this field was organized by Prof. P. Greguss and took place in New York in 1973. A second one was held in Münster in 1976 with the objective to improve the communication among the at that time rather isolatedly working groups in this research domain. The great response to that meeting gave encouragement to the organization of another one in Münster, this time on a more extended international base. Thus, this workshop attracted 85 scientists from 13 countries, i.e. Austria, Brazil, Czechoslovakia, Fed. Rep. of Germany, France, Great Britain, Hungary, Japan, Norway, Sweden, The Netherlands, USA, Yugoslavia.

In order to give a survey of biomedical applications of holography as comprehensive as possible, tutorial review papers gave an introduction to each main session, while special contributions demonstrated the present state of activities in the corresponding field. In addition, techniques related to holographic methods, e.g. Moiré-topography, were included for comparison, and recent developments of special holographic techniques were presented, which may extend holographic applications in medicine and biology, but have not yet been used in these realms, thus far.

This meeting once more demonstrated that the work initiated by the inventor of holography, the Nobel laureate Prof. D. Gabor, who died this year in London on February 9th, is still alive and develops new branches.

Gratefully acknowledged is financial support of this workshop by the Westfälische Wilhelms-Universität, Münster, and the Deutsche Forschungsgemeinschaft, Bonn, as well as the following companies: Agfa-Gevaert, Leverkusen; Carl Baasel, München; Prime Computer, Düsseldorf; Regensbergsche Buchhandlung, Münster; Rohde und Schwarz, Hamburg; Rottenkolber Holo-System, München; Spectra-Physics, Darmstadt; Technitron, München.

Furthermore an address of thanks is directed to all colleagues for administrative and technical assistance in organizing this workshop.

Münster, August 1979   *Gert von Bally*

Contents

Part IX. *Holography in Otology*

Part X. *Acoustical Holography*

Part XI. *Special Holographic Techniques*

I. Introductory Survey

Holography and Its Applications

G.W. Stroke

State University of New York, Stony Brook, NY 11794, USA[1]

Holography, initiated by the work of Dennis Gabor (1) in 1948, has now joined in scope and ramifications the fields of communication and information sciences: It includes, in fact, increasingly the developments of these fields, based on the original contributions of Norbert Wiener and Claude Shannon, among others. Through recent developments in technology, including the laser, as well as in the underlying mathematics and systems theory, including Fourier transform methods, the ramifications of communication and information sciences, as well as of holography, are attaining new levels of importance and interest throughout physics, engineering, astronomy, medicine and biology, among many others. In all contemporary forms, of course, these fields now make increasing use of computers in their several forms. In fact, many of the most impressive applications of holography, beyond its 3-dimensional (3-D) imaging capabilities, and the related remarkable achievements of holographic interferometry, are appearing to be of a 'computational' nature: the field of 'optical computing' is gaining an increasing importance (2), and its potentials were recognized clearly from the beginning when the first textbooks started to appear in the middle 1960's (3)(4)(5) etc. Even earlier, in the late 1950's, holographic principles were used to reconstruct images with coherent light (later lasers) from film records, originally recorded with microwaves, as in the well-known coherent synthetic-aperture side-looking radar (6).

In fact, perhaps the single most important element in the rapid development of holography is the great simplicity that results from the deliberate use of sophisticated but powerful mathematical formulation. To paraphrase Charles H. Townes (7) (co-inventor of the laser, with A.M. Prokhorov and N. Basov), as already mentioned in (3) (page 1), one might say that the recent dramatic developments in holography, and in the related field of electro-optical sciences, including the laser ,

[1] This work was completed during a stay as a Gastprofessor at the Institut für Nachrichtentechnik, Technische Universität, München under a "Humboldt Prize" from the Alexander von Humboldt-Stiftung (1978-1979)

1

"epitomize the great change that has recently come over the character of technological frontiers". The laser, non-linear optics, optical computing, holography, optical image processing, as well as optical communications and integrated optics, to mention only a few, were predicted and worked out "almost entirely on the basis of theoretical ideas of a rather complex and abstract nature". These are not inventions or developments "which could grow out of a basement workshop, or solely from the Edisonian approach of intuitive trial and error". They are creatures of our present scientific age which have come almost entirely from modern theory in physics, communication sciences, and, indeed now holography and electro-optical sciences.

Like Albert Einstein, whose 100th birthday (14 March 1879, in Ulm) is presently being celebrated throughout the world, and whose theories of relativity (1905, 1916) had a number of scientific and, today, well honoured 'precursors', Dennis Gabor, too, (whose life ended in London on 9 February 1979 in his 79th year) had a number of famous 'precursors' when he first described holography in 1948 under title "A New Microscopic Principle", and for which the Nobel Prize in Physics was awarded to him in 1971. Among those, one thinks first of Hans Boersch (8), Professor Emeritus at the Technische Universität-Berlin and of his work (1936-1938), together with that of W.L. Bragg in England, which today, thanks to the most recent developments in holography, has indeed permitted one to obtain images of structures of biological interest, with atomic resolutions, both in electron microscopy, as well as in X-ray crystallography (9) (10). The early pioneering work, which in due course resulted in Dennis Gabor's formulation of what we now call 'holography' (a term proposed by the writer in 1964 to honour Gabor who had coined the world 'hologram' in his 1948 paper, loc. cit.) has been developed and expanded since 1948 by an enormous number of important contributions from universities, industry and scientific institutions throughout the world, too varied and too numerous to be cited in any just form, in full, today. What may be mentioned, however, is that 'holography' seems to have created in the last 15 years or so a surprising 'popular' interest comparable to that created some 40 years ago by Einstein's 'theory of relativity'.

And like the 'theories of relativity', which had to wait almost a full generation, before finding industrially important applications, e.g. in the production of atomic energy (thanks to the famous $E = mc^2$ equation), and also in the laser itself (which had to wait for two further Nobel-prize winning contributions, that of A. Kastler in the optical pumping, and that of N. Basov, A.M. Prokhorov and Ch.H. Townes in the laser design itself), so it seemed that 'holography' (which is still, popularly, most widely known for its famous 3-D imaging capabilities...) would also have to wait perhaps a comparable length of time, in order to demonstrate industrially useful applications of a comparable importance.

For example, like so often in the history of science and technology, one of the first and most surprising applications of 'holography' turned out to be an accidental discovery (11)

2

(12). We speak, of course, of the field of 'holographic interferometry'. When two of the famous 3-D images are superposed
in a single photographic plate (i.e. for example in the latent
image), in such a way that one records a hologram of the same
object in two successive states (e.g. after it has been
slightly displaced, or changed shape between the two exposures
in laser light, for example in the case of an automobile tire
or the wing of an airplane), then one can observe by laser
illumination of the developed holographic plate a 3-D picture
of the object crossed by interference fringes, comparable to
those of classical Mach-Zender interferograms, for example:
these interference fringes represent, like in a geographic
map, the topographic height-differences between the two positions of the 3-D object, and they may be used for establishing mechanical structural properties or for determining
construction defects, among others. Applications to the study
of the living human ear drum (13) have recently also been
demonstrated, among other promising applications of holographic
interferometry in medicine and biology.

Fundamentally, a hologram (which can also be produced artificially with the aid of digital computation, e.g. as in (14)
consists of a diffraction grating, generally produced through
intereference, using laser light, on a photographic plate
which, upon illumination with a beam of laser light, produces
through 'reconstruction' the 3-D (or 2-D) images from the
imaging information stored in the plate. This is quite similar
to the 'reconstruction' of the music stored in a phonograph
record and which may be reconstructed from the encoded record,
by means of a 'pick up' which is made to vibrate by the mechanical deformations of the grooves, as the phonograph rotates
under it. In fact, the similarity between holographic image
storage and reconstruction, on the one hand, and on the other,
phonograph recording and reconstruction of music and sound,
has most recently resulted in the production of phonograph
records using the principles of holography for just that purpose. Both in pictorial and in sound recording, and reproduction, one makes use of the fundamental principle of holography,
namely that of the associative storage of two light waves A and
B, which has the remarkable property that the wave A may be
extracted from the hologram (i.e. reconstructed) by illumination
of the hologram with the wave B, under a wide variety of practically interesting conditions. What is perhaps most remarkable
in this associative principle of holography, is that the wave
B itself may originate from a single point and is, by itself,
sufficient, to produce the wave A which, in turn, may consist
of millions of picture points (i.e. of tens of millions of data
"bits"). This is somewhat like the recall of an entire book,
with thousands of pages, each with hundreds of words, with the
aid of a simple title of the book, with perhaps only a single
word (e.g. "BIBLE"). The ramifications and possibilites of this
associative property of holograms, with its enormous storage
capabilities, are only today becoming slowly appreciated by
non-experts in the fields of data storage and communications,
to its full degree. Completely new types of computers, namely
"optical computers", which operate "in parallel" rather than
"serially", and this in a manner similar to the associative-

3

mode operation of the brain, are already today within the domain of practical possibilities of holography, partly already
demonstrated through scientific examples. Optical computers,
in particular those based on holography in its broadest sense,
could well take on an importance in the presently started
"information age" in the coming decades, comparable to the
"energy age" based on the theories of relativity. The reconstruction of sharp images from blurred photographs (10), (e.g.
in the case of defocusing or vibrations) represents one of the
earliest examples of this type of "optical computing" (see Fig.1)
Many others, particularly applications in X-ray diagnostic
imaging, computerized axial tomography and ultrasonic imaging,
as well as in astronomy, have already shown comparable promise.
Several are presented, in their most recent forms, as a part of
this "International Workshop on Holography in Medicine and
Biology". Others have already been mentioned, at least in their
earliest forms, in detail and in the references both in the
textbooks and in the innumerable 'review articles' which have
been published in the field. Among those, in addition to the
textbooks, and the articles already cited, those of references
(15) and (16) may still be found to be helpful as an initial
introduction.

ACCIDENTALLY BLURRED PHOTOGRAPH

GOOD IMAGE RESTORED BY HOLOGRAPHY

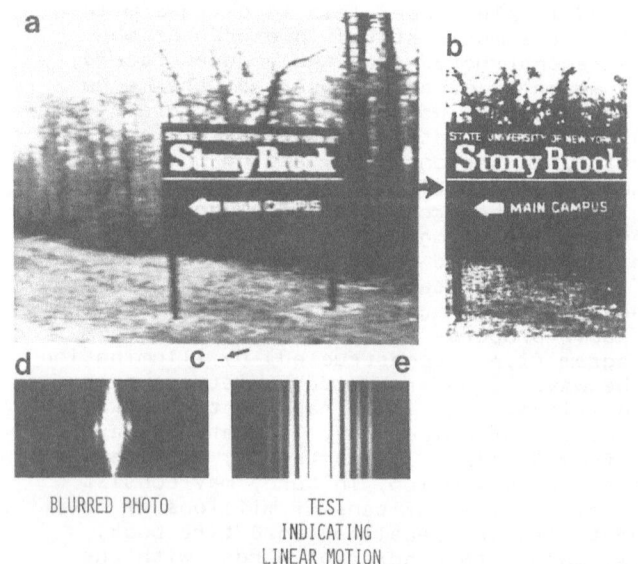

FIG.1. Early example of holographic opto-digital image de-
blurring according to G.W. Stroke, M. Halioua, V. Srinivasan and M. Shinoda, "Retrieval of Good Images from
Accidentally Blurred Photographs", Science, 189 (1975),
261-263. See also Ref. 10 in the text.

In conclusion, one should not forget some promising applications of holography, which, at first sight, seem to be mathematically less complicated, because they involve some extremely remarkable technological elements. We think, in the first place of "holographic zone plates", including "holographic gratings" such as those which recently have already shown promising capabilities in the imaging of living biological specimens, in an imaging "X-ray microscope" using synchrotron radiation (17) and which, in turn, had 'precursors' in the famous work of G. Möllenstedt and co-workers (18) in the 1965 work on the imaging of X-ray stars using a holographic telescope, among others.

References

1 D.Gabor, "A New Microscopic Principle", Nature 161 (1948), 777-778
2 G.W. Stroke, "Optical Computing", IEEE Spectrum 9 (1972, 24-41
  See also:
  H. Platzer und K. Etschberger, "Fouriertransformation zweidimensionaler Signale", Laser + Elektro-Optik 4 (1972), H.1. 39-45, H.2. 43-49
3 G.W. Stroke, An Introduction to Coherent Optics and Holography (New York, Academic Press: First Edition, 1966, Second Edition, 1969)
4 J.W. Goodman, Introduction to Fourier Optics (San Francisco, Mc Graw-Hill Book Co.: 1968)
5 H. Kiemle und D. Röss, Einführung in die Technik der Holographie (Franfurt/Main, Akademische Verlagsgesellschaft: 1969)
6 L.J. Cutrona, E.N. Leith, L.J. Porcello and E.E. Vivian, "On the Application of Coherent Optical Processing Techniques to Synthetic-Aperture Radar", IEEE Proc. 54 (1966), 1062-1032
7 Ch.H. Townes, in "The Age of Electronics" (C.F.J. Overhage, ed.), p.166 (New York, Mc Graw-Hill Book Co.: 1962)
8 H. Boersch, "Zur Bilderzeugung im Mikroskop", Z. Techn. Physik 19 (1938) Nr. 2, pp. 337 ff.
9 G.W. Strocke and M. Halioua, "Three-Dimensional Reconstruction in X-Ray Crystallography and Electron Microscopy by Reduction to Two-Dimensional Holographic Implementation", Trans. Amer. Crystall. Assoc. 12 (1976) 27-41
10 G.W. Stroke, M.Halioua, F. Thon and D. Willasch, "Image Improvement and Three-Dimensional Reconstruction Using Holographic Image Processing", IEEE Proc. 65 (1977) 39-62
  See also:
  G.W. Stroke, M. Halioua, R. Sarma and V. Srinivasan, "Imaging of Atoms: Three-Dimensional Molecular Structure Reconstruction Using Opto-Digital Computing", IEEE Proc. 65 (1977) 589-591
11 K.A. Stetson and R.L. Powell, "Hologram Interferometry", J. Opt. Soc. Am. 56 (1966) 1161-1166
12 G.M. Brown, R.M. Grant and G.W. Stroke , "Theory of Holographic Interferometry", J. Acoust, Soc. Am. 45 (1969), 1166-1179 (This paper contains also a history of the field, with complete references)
13 G. von Bally: see article in this volume (which also contains history and references)
14 A.W. Lohmann and D.P. Paris," Binary Fraunhofer Holograms Generated by Computer", Appl. Opt. 6 (1967) 1739-1748
15 G.W. Stroke, "Holography" in The Science Teacher, Vol. 34, No. 7 (October 1967), (8 pages)
16 D. Gabor, W.E. Kock and G.W. Stroke, "Holography", Science 173 (1971), 11-23

17 G. Schmahl, D. Rudolph and B. Niemann, "X-Ray Microscopy of
   Biological Specimens", Jour. de Physique, 39 (Juillet 1978),
   C4-202 (Colloque C4:Supplement au No. 7)
18 H.J. Einighammer, "Zur Abbildung von Röntgensternen mit dem Hologram-
   Teleskop", Optik 23 (1965-1966), Nr. 7, pp. 627-641

II. Holography in Orthopedics

Holography in Orthopedics

K. Piwernetz

Institut für Medizinische Optik der Universität München
Theresienstr. 37, D-8000 München 2, Fed. Rep. of Germany

G. von Bally

Hals-Nasen-Ohrenklinik der Westfälischen Wilhelms-Universität
Kardinal-von-Galen Ring 10, D-4400 Münster, Fed. Rep. of Germany

1. Introduction

In the field of technical applications holography is a fully established method [1]. Recent developments show that it increasingly
gains also in importance as a technique for nondestructive testing
in biomedical research.

Although holography has been introduced into orthopedics already in 1971 [2], there are still only few publications in this
field up to now. Thus, in this paper not only applications are
reviewed but also some valuable techniques are summerized that
have not yet been applied in this speciality. Emphasis will be put
on the advantages and experimental problems of holographic applications with special regard to orthopedic demands.

Three fields will be considered:

3-D imagery

surface contouring

deformation analysis.

Holographic image processing techniques are omitted because basi-
cally no orthopedic problems will be solved using these methods
but primarily radiographical ones [3].

Figure 1 gives a survey of that parts of the skeleton which
already have been examined by means of holographic techniques.

2. 3-D Imagery

The application of holography to record 3-D images or to display
the topography of irregular shaped surfaces has not yet found the
same great interest as the determination of surface deformations
under various loads by means of holographic interferometry.

Thus, in the field of pure 3-D imagery no orthopedical applications of holography could be found in literature. A reason for

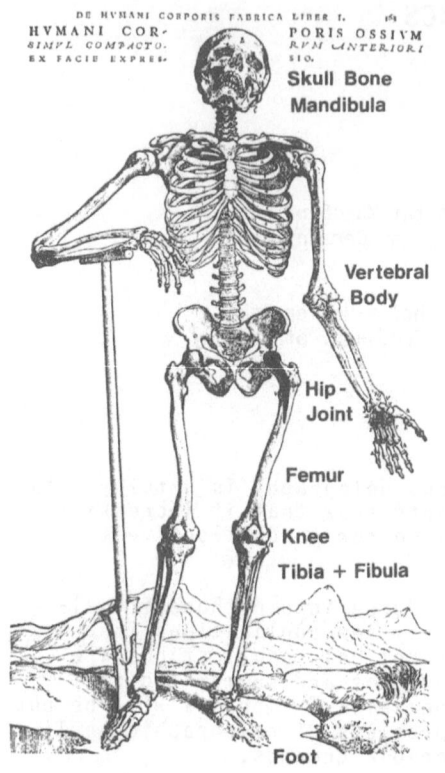

DE HVMANI CORPORIS FABRICA LIBER I. 163
HVMANI COR-   PORIS OSSIVM
SIMPL COMPACTO- RVM ANTERIORI
EX FACIE EXPRES- SIO.

Skull Bone

Mandibula

Vertebral
Body

Hip-
Joint

Femur

Knee

Tibia + Fibula

Foot

Fig.1 Holography in orthopedics. Parts of the skeleton that have been examined holographically up to now [after A. VESAL (1543)].

this fact may be that the advantages of this method do not balance the costs and the complexity of the recording technique, that up to now has to be performed in a specially equipped laboratory.

Since the structures interesting for clinical orthopedists are not directly visible, holographic investigations using ultrasonic waves with the capability of penetrating tissue have been carried out. Biomedical applications of acoustical holography are reported on elsewhere in this issue [4], and therefore not described in detail here.

Much work, however, has been done to improve indirect methods of 3-D imagery in radiographical diagnosis, which is important for instance for the quantification of anomalies in orthopedics. Two different holographic concepts have to be mentioned in this field: the coded aperture method and the synthesis of tomograms. An introduction to these techniques can be found in [5].

3. Surface Contouring

In order to display the topography of irregular shaped surfaces two different concepts have been introduced:

Moiré topography and

holographic topography.

It may be better to distinguish between incoherent and coherent topography, since in principle the holographic techniques can be understood as Moiré patterns, too [6].

Medical applications of incoherent surface contouring are described e.g. in [7]. For comparison in this chapter some advantages of coherent methods are pointed out (Fig. 2).

Using coherent techniques the surface itself and the topographical pattern are stored three dimensional. The first two methods mentioned in Fig. 2 are easy to perform with regard to both the optical setup as well as the recording procedure , but the more the sensitivity vector is turned to the direction of observation the more the patterns suffer from projection shadows. Taking no account of the draw-back of the two-step exposure, sandwich-contouring [8] renders the unique possibility to vary the sensitivity vector and the sensitivity itself over a wide range during the reconstruction. The sensitivity of these three techniques is comparable to that of Moiré methods.

Both multiple frequency and multiple refractive index holographic contouring are sensitive to path differences within the direction of observation. The depth resolution is controlled only by optical parameters without any change of the geometrical arrangement, i.e. this technique can operate through small apertures.

A disadvantage is that real-time operation cannot be performed as easy using the latter method as using Moiré techniques.

Method	Advantages	Problems
Change of Illumination Direction	Simple Setup	Shadow 2 - Directions
Fringe Projection	Real - Time	
Sandwich - Contouring	Simple Setup Variable Sensitivity after Recording	Recording Procedure
Multiple Refractive Index - Contouring	Wide Continuous Range of Sensitivity Small Aperture	Extreme Stability Object Must be Placed in Tank
Multiple Frequency Contouring	Highest Sensitivity Small Aperture	Most Complicated (Chrom. Aberrations, Adjustment 2 Frequency - Laser)

Fig. 2 Holographic surface contouring techniques

Additionally, holographic contouring is more expensive and more complicated. Except sandwich-contouring these techniques are explained in detail in [1].

In clinical orthopedics normally no high sensitivity is necessary, but real-time operation is very important. Therefore Moiré topography is dominating. In basic research often higher resolution and sensitivity is required, e.g. in measurements of wear of implant materials. Thus, in the latter case the topography of an artificial knee-joint before and after being subjected to wear has been recorded using multiple refractive index holographic contouring [9].

4. Deformation Analysis

4.1 General Considerations

Deformation analysis is the by far most predominant application of holography in orthopedics [10]. In general the objects are very complex both in surface structure as well as in architecture. Since they often deform in a complicated manner, pointwise contactive mechanical measurements turn out to be cumbersome.

Holographic interferometry gives a detailed survey of surface displacements at one glance. Although the reconstruction of one interferogram shows at a fixed direction of observation only one component of the deformation vector, it contains in principle the complete information to evaluate all three Carthesian components of this vector. But the determination of these vectorial components with a sufficient accuracy is often not elementary [11], [12].

Before considering particular techniques the general advantages and some problems of holographic interferometry are discussed. The advantages are listed in Fig. 3.

1. Measurement of the Object itself
2. Contactless Measurement
3. 2-Dimensional Display of the Deformation Pattern
4. High Sensitivity

Fig.3 Advantages of holographic interferometry

1. The measurements can be carried out on the object itself. No models are required with the demand for mechanical equivalence which often cannot be guaranteed, especially not in case of orthopedic objects. Therefore, in principle all parameters that influence the deformational behaviour remain accessible. Primarily anisotropic effects and complex mechanical interrelationships have to be taken into consideration in this context.

2. No mechanical contact is involved in the measuring procedure. Thus, the boundary conditions that are of great importance for the deformational behaviour are determined only by the way how the object is supported and how the various stresses are introduced.

3. The deformation pattern is displayed directly over the illuminated surface. Areas of high strain for instance can be seen immediately and need not to be evaluated labourously. Moreover one can decide at once whether the object has been supported stable or has suffered from additional rigid body motions.

4. The sensitivity depends on the geometrical arrangement and is in the order of magnitude of the wavelength of the laser light.

Additionally, holographic interferograms are easy to record if only care is taken to meet the interferometric stability requirements(i.e. parasitic motions have to be less than one tenth of the used wavelength) for all components of the set-up, including the object . This may sometimes be a problem, especially when irregular shaped orthopedic objects with specific boundary conditions are concerned. Thus, superimposed rigid body motions shall not or even cannot be avoided. In these cases there are three possibilities to obtain the pure deformation pattern:

1. In real-time interferometry the motion can be compensated by manipulating the wavefront of the illumination and/or the reference beam as shown in [13].

2. Sandwich-holography solves this problem in cases where rigid body motions are inherent in the process under consideration or where the amount of these motions is by far higher than the deformation itself. This technique was developed for use of CW-lasers [14] as well as pulsed lasers in [15].

3. If rigid body motions occur only in particular cases or if the interferogram cannot be recorded again, the motion can be compensated by means of a Moiré-technique [16].

Methods like these are known as fringe-control techniques.

Solving the problem to avoid rigid body motions the possibility of contactless measurements should not be depreciated by introducing new intricate boundary conditions through a stable but complex fixation of the object on the support. With regard to the analysis of the deformations it is always profitable to adapt the deforming stresses and the corresponding boundary conditions to the architecture and the mechanical properties of the orthopedic object in a cooperative way. That means, either one will try to find boundary conditions that can be described mathematically, if technical properties are of interest, or one will try to simulate the anatomical realities, if the deformation behaviour shall be investigated only qualitatively. This simulation is especially difficult, when an approximation of the force distribution by the various ligaments and tendons shall be achieved. As numerous the problems are as numerous are their solutions. Thus, recommendations can only be very general and no limits are set to the ingenuity of the experimentators. Success or failure of an experiment often are caused thereby.

Thus far only the recording procedure has been taken into consideration, but the quantitative evaluation is as important as well . Since this problem can only be mentioned in this article, it is referred to comprehensive literature on this subject [1], [11], [12], [17].

4.2 Applications

The applications described in the following chapters are classified with respect to the corresponding underlying holographic technique. Before reporting in particular on these applications the different techniques are listed in Fig. 4.

Method		Range of Application
Real - Time		Static and Dynamic Loads
		Continuous Recording
Double - Exposure	Pulse-Laser	Rapid Deformations
		Superimposed Rigid Body Motions
	C W - Laser	Static Loads

Fig.4 Comparison of different holographic interferometric techniques used in orthopedics

4.2.1 Double Exposure Interferometry

One will hardly succeed in detecting the mechanics of complex frameworks of bones not applying holographic interferometry. The double exposure technique was used to study the interactions of the different parts within the tibia-fibula system [18]. Similarly the complex mechanical action of the ankle-joint has been analysed [19]. Measuring static deformations of fresh and macerated human vertebral bodies shall help to determine its elastomechanical properties in more detail [20], [21], [22].

These measurements have been carried out using CW-lasers. Double exposure interferometry with such lasers requires the object to be fixed stable on the support and the deformations to be static. As a compensation for these restrictions only a low-output-power is needed.

In order to study fast dynamic deformations by double-exposure interferometry pulsed-ruby-lasers are used. Due to the short duration of the pulses the objects need not rest absolutely stable on a support. How fast the deformations have to or can be, is determined by the available separation time of the two pulses that form the interferogram.

Interferometry with a pulsed-laser was applied to investigate the deformations of vertebral bodies and intervertebral discs under axial dynamic loading [23], [24]. A ruby laser has been triggered by a piezoelectric cell that measured the rapid changes of the compressive force (about $2.5 \cdot 10^5$ N/sec).

The visualization of chest motions is a good example to show that in double-exposure interferometry with a pulsed-laser the objects not always need to be fixed on a support [25].

4.2.2 Real-Time Interferometry

Real-time interferometry renders the unique possibility immediately to see deformations rise under increasing load. Due to the high sensitivity the object together with the whole set-up has to rest during the observation time interferometrically stable. The recording procedure of the zero-hologram is substantially simplified using thermoplastic recording media which overcome wet development of conventional photographic plates.

These media were applied to comparative investigations on the mechanical influence of two types of fracture-fixation plates on the stability of a fractured human tibia [26]. Similarly, real-time observation helps in studying the different stress- and strain-distributions that occur when dried human tibiae are fractured at different angles to the axis and are supplied with the same fixation plate [27]. The interferograms show at once areas of high strain, what can be important for successful treatment.

Another question of great clinical importance which may be answered by means of holographic interferometry is how different artificial hip-joints influence the strain distribution in a femur under axial compressive load [26]. In order to compensate for the relatively large motions that superimpose on the local deformations of such an object, real-time interferometry can advantageously be combined with fringe-control techniques [28].

5. Conclusions

Although various results demonstrate the feasibility of holographic interferometry in orthopedics many problems are still open in this special field of applications. The capabilities of this method are convincing but they are in conflict with the technical and experimental expenditure that is necessary before the first hologram can be taken.

6. References

1 R.K. Erf, Holographic Nondestructive Testing, Academic Press, New York and London (1974)
2 U. Hanser, Z. Orthop. 110, 871 (1972)
3 P.T. Gough and R.H. Bates, Comput. Biomed. Res. 5, 700 (1972)
4 P. Greguss, this issue
5 H. Weiss, this issue
6 N. Abramson, Optik 37, 337 (1973)
7 H. Takasaki, this issue
8 N. Abramson, Appl. Optics 15, 200 (1976)
9 M.J. Lalor, J. Groves and J.T. Atkinson, this issue
10 M. Hoke and G.v.Bally (eds.), Proc. Symp. 1976 Spec. Res. Area 88 and Conf. on Electrocochleography and Holography in Medicine, Münster, Vol. I (1976)

11   H. Kohler, Optik 47, 469 (1977)
12   L. Ek and K. Biedermann, Appl. Opt. 16, 2535 (1977)
13   G. Ferrano and G. Häusler, this issue
14   N. Abramson, Appl. Opt. 13, 2019 (1974)
15   H. Bjelkhagen, Appl. Opt. 16, 1727 (1977)
16   K. Piwernetz, Opt. Acta 24, 201 (1977)
17   R.J. Collier, C. Burckhardt, L.H. Lin, Optical Holography,
     Academic Press, New York and London (1971)
18   J. Wagner, J. Ebbeni, M. Clemens, Acta Orthop. Belg.(Suppl.)
     41, 24 (1975)
19   D. Vukicevic, V. Nikolic, J. Hancevic, S. Vukicevic, see [10],
     pp. 333
20   Th. Wesendahl, K. Piwernetz, G.v.Bally, J. Polster, Laser and
     Electro-Optic 1, 37 (1977)
21   K. Piwernetz, see [10], pp. 321
22   K. Piwernetz und R. Röhler, Frühjahrsschule 1978: Holografische
     Interferometrie in Technik und Medizin, Hannover (1978)
23   Th. Wesendahl und G.v.Bally, Frühjahrsschule 1978: Hologra-
     Interferometrie in Technik und Medizin, Hannover (1978)
24   Th. Wesendahl und G.v.Bally, see [10], pp. 327
25   S.M. Zivi, G.H. Humberstone, Med. Res. Eng. 9, 5 (1970)
26   U. Hanser, see [10], pp. 343
27   K. Hardinge, G.R. Bremble and M.J. Lalor, see [10], pp. 307
28   G. Häusler, T. Schwenk und K. Seidel, see [10], pp. 349

Elastomechanical Properties of Trabecular Bone from the Human Vertebral Body

K. Piwernetz and R. Röhler

Institut für Medizinische Optik der Universität München
Theresienstr. 37, D-8000 München 2, Fed. Rep. of Germany

1. Introduction

Literature shows only few indications on mechanical anisotropy
of trabecular bone. In detail compressive anisotropy is repor-
ted upon the trabecular bone of a patella |1|. As for bone from
the vertebral body the compressive strength depends essentially
on the direction of the load |2|,|3|. The variations in the com-
pressive strength have been discussed on the base of the Euler
formulas |4|. Another paper postulates trabecular bone to be
isotropic |5|.

Up to now the elastomechanical constants have not yet been
measured in particular. They are assumed to give a more detailed
understanding of the mechanical behaviour of the entire vertebral
body.

This paper deals with the determination of the compressive
and shearing stiffnesses in three directions by means of holographic interferometry. In comparison the stiffness is given
for torsional load around the vertical axis. These results complete earlier presentations about the compressive behaviour of
the entire vertebral body |6|.

2. Preparation of the Bones and Measuring Procedure

The blocks of trabecular bone were obtained from one single
vertebral body. Care was taken that the bordering areas were
exact rectangles. The blocks have been fixed on the support by
a two-component bone cement (Palacos) to get comparable boundary
conditions in all cases. The various stiffnesses were determined
in the different directions consecutively.

The ultimate strain has been determined by compression tests
to be about 1%. Thus, only small deformations can prevent the
fine trabeculae in the interior from microfractures. In order to
remain within the range of pure elastic deformations the displacements have to be kept very small, too.

This condition can be fulfilled due to the high sensitivity of
holographic interferometry. Additionally, the linear deformation
behaviour can be controlled by the 2-D representation of the
displacement field and the influence of the boundary conditions

roughly be estimated. From the deformation patterns the applicability of St. Venant's principle was concluded.
The optical setup (Fig.1) is arranged such that the directions
of the displacement vectors coincide with the line of sight in
shearing and torsional tests or are perpendicular to it in compression tests. This is of importance with regard to simplicity
and accuracy of the evaluation.

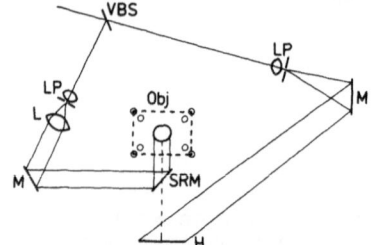

Fig.1   Optical Setup (Obj.: object,
H: hologram, VBS: variable beam
splitter, SRM: semi-reflecting mirror, M: mirror, L: lens, LP: lens-
pinhole combination)

In the first case the interference fringes localize near the
surface of the object and at infinity in the second one |7|.
Therefore two different evaluation methods have to be applied:
static evaluation by counting fringes in the reproduced inter-
gram by means of microdensitometry as shown in Fig.2 and dynamic evaluation by changing the direction of the observation
vector |8|.

3. Results

The results were obtained on the base of Hooke's law, that means
bone was assumed to be homogeneous. This assumption seems justi-
fied as 1. the distances between the single trabeculae are small
compared to the external geometrical dimensions and 2. the deformation patterns are similar to that homogeneous bodies would
show under the same conditions (conf. Fig.2).

To give an idea of the strains that have been measured: under
axial compression strains were found to be in the order of magnitude of about 10^{-3} and under shearing forces of about 10^{-4}.
These results demonstrate the different sensitivities of the
dynamic and static evaluation methods and show that the average
strains are one order of magnitude smaller than the ultimate
strain, so that one presupposition of Hooke's law - small deformations - certainly seems to be fulfilled.

The compressive stiffnesses turned out to be essentially anisotropic. The values in the three main directions - axial (z),
lateral (y), and ant.-post. (x) - are given in Fig.3. The highest
stiffness was measured in the axial direction, whereas both others
correspond with one another within the accuracy of the measurement. Their values are less than one third of the axial one.

A deformation perpendicular to the direction of the compressive
force has not been found within the sensitivity of the measurement.

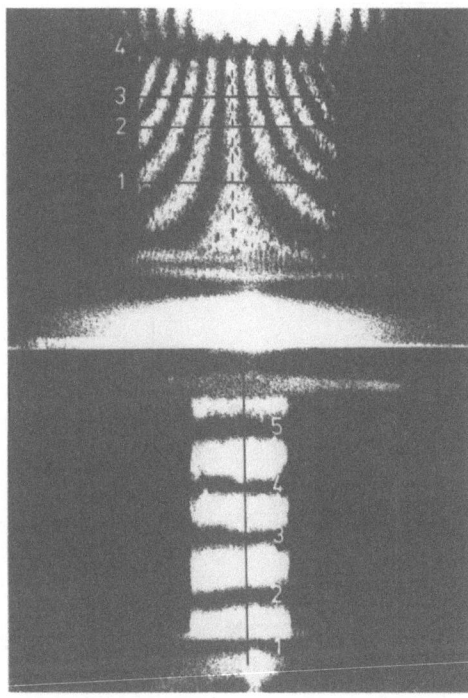 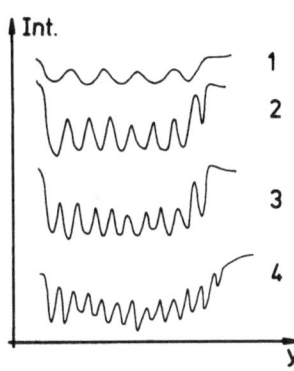

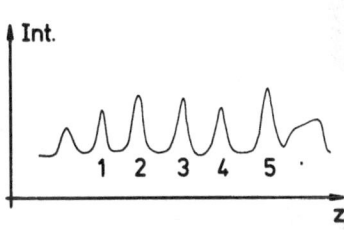

Fig.2 Quantitative Evaluation of the Deformation Patterns
In the upper part a block of spongy bone is subjected to an
angular momentum at the top and is fixed on the bottom. In
the lower half another block suffers from a shearing force
perpendicular to the plane of the drawing. The right side
shows the intensity distribution along the microdensitometer-
traces (solid lines in the left figures).

Applying a shearing momentum the blocks deform in a combined
bending and shearing manner. This has to be marked for getting
proper results. The shearing stiffnesses in the directions perpendicular to the axis (S_{zx} and S_{zy}) equal one another and
are nearly comparable to the value obtained by the torsion around
this axis (G_z), according to the theory for homogeneous
bodies. The values are shown in particular in Fig.4.

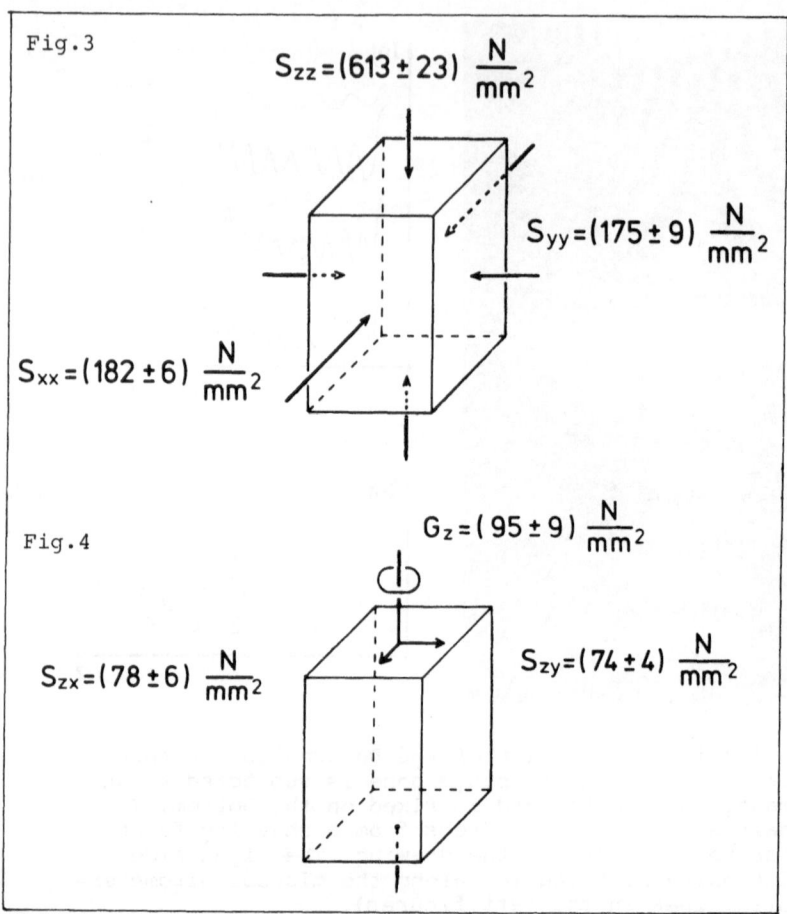

Fig.3 $S_{zz} = (613 \pm 23) \dfrac{N}{mm^2}$

$S_{yy} = (175 \pm 9) \dfrac{N}{mm^2}$

$S_{xx} = (182 \pm 6) \dfrac{N}{mm^2}$

Fig.4 $G_z = (95 \pm 9) \dfrac{N}{mm^2}$

$S_{zx} = (78 \pm 6) \dfrac{N}{mm^2}$

$S_{zy} = (74 \pm 4) \dfrac{N}{mm^2}$

Fig.3 Compressive Stiffnesses of Trabecular Bone from the Human Vertebral Body
(x: anterior-posterior, y: lateral, and z: superior-inferior)

Fig.4 Stiffnesses According to Shearing (S_{zx} and S_{zy}) and Torsional Loading (G_z)

4. Discussion

The anisotropy of the compressive stiffnesses supports the theory of the functional adaptation of bones to external loads they have to bear in normal function |9|. The highest load has to be transmitted in axial direction that is the direction in which the body weight stresses the vertebral body.

The different compressive stiffnesses can be understood from a model of the anatomy of the trabecular bone |4| that gives the background for an approximative interpretation of the elastic behaviour by the Euler formulas. The vertical trabeculae are subjected to a high normal stress and show - according to the feedback model of the built-up of bones |9| - larger diameters whereas the radial ones only have to support them against lateral bending and are for the same reason more slender. This difference in diameter is a reason for the different stiffnesses, perhaps modified by the density of the vertical and radial trabeculae.

From this anatomical point of view it similar can be understood why the shearing stiffnesses do not show the same distinct anisotropy. In what direction the lattice is ever cut the rectangles of trabeculae that are sheared always contain at least two opposite sides built up by the more slender trabeculae that in this way will determine the shearing stiffnesses.

The model mentioned above is supported by the fact that no lateral deformation has been found in the compression tests.

With regard to the complete vertebral body it can be concluded that the deformations of the cortical bone are due to a deformation of the cortical bone itself. The amount of the deformation can quantitatively be explained by a motion of the corticalis against the radial compressive stiffness of the trabecular bone in the interior.

5. References

1   P.R.Townsend, P.Raux, R.E.Miegel, and E.L.Radin, J.Biomech. 8,363 (1975)
2   J.Galante, W.Rostoker, and R.D.Ray, Calcif.Tiss.Res. 5, 236 (1970)
3   R.Plaue and H.Roesler, Z.Orthop. 110,582,(1972)
4   G.H.Bell, O.Dunbar, J.S.Beck, and A.Gibb, Calcif.Tiss.Res. 1,75 (1967)
5   J.McElhaney and V.Roberts, AIAA 9th Aerospace Science Meeting, New York, N.Y., Jan 25-27, (1971)
6   K.Piwernetz in: M.Hoke and G.von Bally (eds.): Proc. Symp. 1976, Spec. Res. Area 88 and Conf. on Electrocochleography and Holography in Medicine, Münster (1976)
7   R.J.Collier, C.B.Burckhardt, and L.H.Lin, Optical Holography, Academic Press, New York and London (1971)
8   H.Kohler, Optik 47,135 (1977)
9   H.M.Frost, Bone Modeling and Skeletal Modeling Errors, in: C.C.Thomas (ed.), Orthopedic Lectures, Vol.IV (1973)

Holographic Studies of Wear in Implant Materials and Devices

M.J. Lalor, D. Groves, and J.T. Atkinson
Department of Mechanical, Marine, and Production Engineering,
Liverpool Polytechnic, Byrom Street, Liverpool, U.K.

1. Introduction

Wear is a very significant problem in the performance and design of total replacement prostheses. In recent years a great deal of interest has been shown by many workers in the measurement of wear in bioengineering materials and devices. Most of the experimental work has been done using the standard wear test devices such as pin on disc and reciprocating machines. The method of wear measurement in these experiments has nearly always been gravimetric. The major disadvantage in using these standard test methods is that the wear machines will not reproduce in vivo results since the test conditions are totally different. To overcome this difficulty joint simulators are being designed and built by various workers [1, 2]. For these simulators to yield useful findings new methods for measuring wear must be developed as the gravimetric methods in common use are no longer applicable since:

(a) They are not sensitive enough.
(b) They do not furnish the user with information about the location of the wear and this information is essential when comparing different designs of prostheses.

Any new method of wear measurement must satisfy the requirements of

(1) Very high sensitivity.
(2) Wide range of application.
(3) Providing information about the location of wear.
(4) Being usable on prostheses worn in vivo, as well as on wear test samples.

The authors consider optical contouring to be a significant solution to the problem of wear measurement of simulator worn prostheses and in vivo worn prostheses after removal from patients. It must be stressed here that by 'wear' in this context is meant the change in volume caused by use, which of course for plastic materials is not only due to loss of material.

2. Optical Contouring

2.1 Dual Index Holographic Contouring (DIHC) is carried out by sequential immersion of the object or specimen in transparent liquids of refractive indices n_1 and n_2, two holographic exposures of the object are made through a window in the liquid tank, before and after the liquid is changed. When the object beam is reconstructed, interference fringes denoting loci of constant height from the reference plane (i.e. the window) will be seen on the

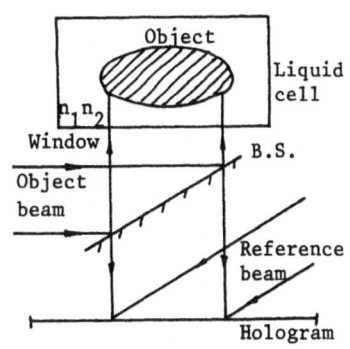

Fig.1

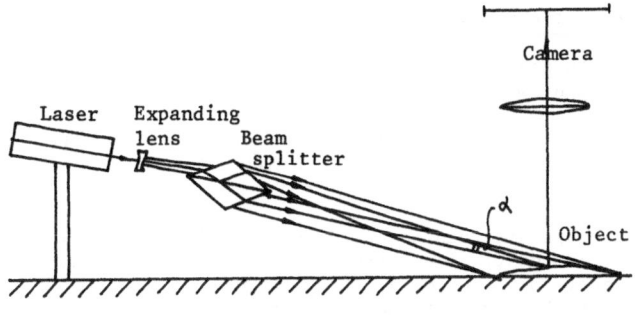

Fig.2

object's surface. The height difference between two adjacent fringes, Δh, called the contour depth, will be

$$\Delta h \simeq \lambda / \left(2|n_1 - n_2| \right)$$

where λ is the laser wavelength used [3].

The contour depth can be varied typically between 10 μm and several metres, so that engineering components may be contoured at meaningful contour depths and accurate surface metrology carried out. DIHC can be used to make maps of abrading components before and after wear; examination of these maps will give quantitative information on the distribution and amount of wear in situations where conventional wear measurement is not possible. A typical contouring arrangement is shown in Fig.1.

2.2 Dual Source Contouring (DSC). A schematic diagram of the optical arrangement is shown in Fig.2. The laser beams emergent from the beam splitter combine to form a fringe pattern on the sample surface. This fringe pattern can be considered as a contour map of the sample. The contour depth will be given by

$$\Delta h \approx \frac{\lambda}{\alpha}, \quad \text{where } \alpha = \text{the small angle between the beams.}$$

Other optical methods of contouring are available, but the above two are the simplest and least expensive.

The depth and volume of wear can be estimated in at least two ways; both these methods are discussed for the first example, given below.

3. Examples

3.1 In Vivo Worn Freeman Swanson Knee Prosthesis (DIHC)

Figures 3 and 4 are contour maps of an unworn and (different) worn tibial component of a Freeman Swanson Knee Prosthesis; $\Delta h = 165$ μm and 190 μm respectively. The scale is approximately 0.5.

The unworn component map shows an error of form of about $\Delta h/2$ (= 82 μm) on the extreme right of the component.

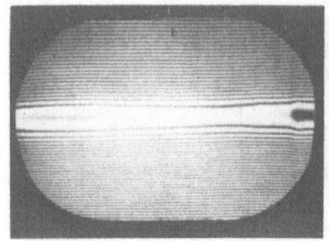 

Fig.3   Fig.4

The worn component map shows the area of wear; this area can be split into two sections (as shown). One section shows mild wear to a depth of ∿ 30 μm. The other shows more severe wear, up to 600 μm in places. This worn component was used by a light to moderate user for about 2 years.

The wear can be estimated in two ways:

(i) The pre-wear surface can be assumed (if the surface has a simple geometry) and an assumed pre-wear map drawn.

(ii) If the pre-wear surface is highly convoluted (for example a tooth), the pre-wear map cannot be assumed (in general) and must be made prior to the test.

The pre-wear and post-wear maps are then compared in one of two ways. If the contour depth in each map is identical and the contours are drawn relative to the same reference plane, then the moiré pattern formed by superimposing the two maps will correspond to a map of the worn surface. In practice it is difficult using DIHC to reproduce the contour depth exactly.

The pre- and post-wear maps can also be compared by digitising the contour information and computer processing. This processing is very accurate, and will be discussed in a later section.

3.2 In Vivo Worn Manchester Knee (DIHC)

Figures 5 and 6 show contour maps (actual size) of the medial section of a tibial component of a Manchester knee design (left) worn in vivo for about two years. The contour depths are 113 μm and 55 μm respectively. The lateral component was subject to catastrophic wear, thought to be caused by the presence of bone fragments, and is not shown. The wear scar is clearly visible, the volume of wear can be calculated as 15 mm^3 [4] by assuming the line drawn between A and B is a pre-wear contour of the surface; the depth of wear at C is therefore approximately 150 μm (3 × Δh). The wear scar was assumed to be elliptical in area and the volume found in this case by hand calculation.

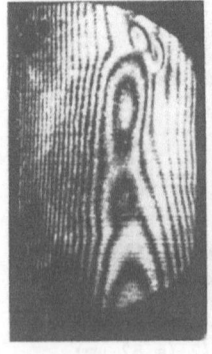 

Fig.5   Fig.6

22

3.3 Acetabular Cups (DIHC)

It is possible to measure wear in hip prostheses using DIHC. Figure 7 is

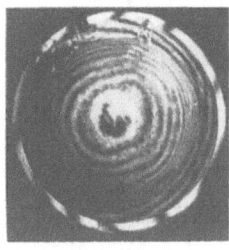

an example of a map of a worn acetabular cup. The scale is 0.5 and the contour depth is 1.5 mm. Wear measurement of simulator worn and in vivo worn acetabular cups is in progress. Because of the steep slopes of the sides of these cups, which results in the contour fringes being very close together if the cup is placed symmetrically towards the reference plane, it is necessary to take several contour holograms with the cup at different orientations and use the computer to build up an overall picture.

Fig.7

3.4 Application of the Technique to Dental Materials

Work on finding the wear rates of the suggested alternatives to dental amalgams [5] is of prime importance at present. Most of this work has been performed using reciprocating devices, despite the sophistication of some of these machines they do not simulate oral conditions. Ideally the in vivo wear rates of these materials should be found

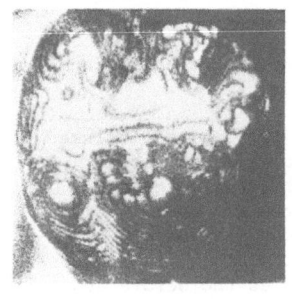

and compared. The authors are working on a programme leading to the in vivo comparison of dental filling materials, this work has so far included the development of a Dual Source Contouring (DSC) system for measuring wear on amalgam samples [6]; a natural extension of this work has been mapping of real teeth using DIHC. A proposed programme of work, leading to in vivo wear measurements on dental restoratives, is:

(i) Dentures would be filled and contoured at regular intervals. A major difficulty here is that the wear pattern measured may not correspond to the in vivo case since the chewing action is different. This approach is easier, possibly more realistic and certainly cheaper than building a simulator.

Fig.8

(ii) Impressions of filled teeth would be made at regular intervals. The impression would then be contoured. This approach is rather impractical, since the shrinkage of the impression is an unknown factor.

(iii) Following the development of a suitable dual frequency holographic contouring [7] arrangement using a pulsed laser, filled teeth would be contoured in vivo at (say) 3 monthly intervals. The design and manufacture of a specialised instrument using fibre optics is a long term objective. Figure 8 is an example of a contour map of a human molar; $\Delta h \simeq 200$ μm.

3.5 In Vivo Worn Mitral Heart Valve (DSC)

Figure 9 shows contour maps of an in vivo worn prosthetic mitral heart valve made using the DSC arrangement shown in Fig.2. The wear scar is easily visible; as indicated the volume of wear can be calculated as ~ 0.4 mm^3 [4]. The valve was used for 2 years. The scale is 1.5.

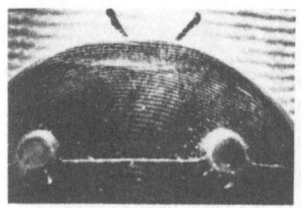  

Δh = 64 μm   Δh = 95 μm   Δh = 288 μm

Fig.9

3.6 Simulated Wear Scar (DSC)

Figure 10 shows maps of a simulated wear scar in a metal block. The maximum depth of wear is easily found as ∿ 750 μm. Similar maps are found when reciprocator test samples are encountered. The scale is 1.

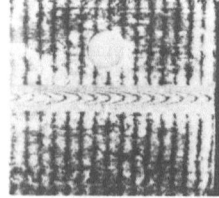 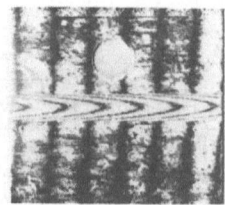

Δh = 115 μm   Δh = 288 μm

Fig.10

4. Computer Analysis

A contour map of a surface contains three dimensional information about that surface. Hence the volume bounded by that surface and a reference plane parallel to the planes on which all the contours lie can be estimated.

To determine the volume change of an object, say due to wear, the volume bounded by the before wear surface and the reference plane is subtracted from the volume bounded by the after wear surface and the same reference plane. Such volume estimations from contour maps requires a computer as in general the contours are not described by simple equations and so must be recorded digitally. A computer is required to perform repeatedly laborious numerical methods to estimate volumes. The computer scales the before and after wear contour maps and makes the two images coincident, and overcomes the problem of having to exactly reorientate the object after wear to its before wear orientation in the contouring system.

4.1 Digitisation

A digitiser consists of a table upon which a photograph of the contour map is taped. A handrest with 'aiming' crosswires can be placed at any point on the photograph and the two dimensional coordinates of that point recorded on paper tape. If each contour is followed using the handrest and points taken close enough for the contour to be well approximated by the curve formed by joining all adjacent points with straight lines, and the height from the reference plane of each contour recorded on the paper tape, the three dimensional information of the contour map is recorded on the computer readable punch tape.

4.2 Numerical Methods

The volume bounded by the surface represented by a contour map and the reference plane can be estimated by determining areas projected in the direction

24

of view (orthogonal to the reference plane) between adjacent contours and applying numerical methods which involve the projected areas and the heights of each adjacent pairs of contours. The projected areas are determined from the digital information on paper tape using the Trapezium rule.

The reference plane is assigned index number i=o. The plane one contour depth Δh above that is given index number i=1, etc.

4.2.1 Open contours.

The volume bounded between two parallel open contours and the reference plane is

$$\delta V_o = \Delta h \left[i + \tfrac{1}{2} \right] A_i$$

where Δh is the contour depth

　　　i is the index of the lower of the two adjacent open contours

　　　A_i is the projected area orthogonal to the reference plane between contours i and i+1.

4.2.2 Closed contours.

The volume bounded between two circular adjacent closed contours is

$$\delta V_c = \frac{\Delta h}{3} \left[A_i + A_{i+1} + \sqrt{A_i A_{i+1}} \right]$$

where Δh is the contour depth

　　　i is the index number of the lower contour

　　i+1 is the index number of the upper contour

　　　A_1 is the area projected orthogonally to the reference plane by the lower contour

　　A_{i+1} is the area projected orthogonally to the reference plane by the higher contour

The volume bound between a surface represented by open contours and the reference plane is the summation of all open contour volume elements.

$$V_o = \sum_{\substack{\text{All } i \\ \text{open}}} \Delta h \left[i + \tfrac{1}{2} \right] A_i$$

If any closed contour appears on the surface the volume of the mound or depression represented can be determined by summation of the closed contour volume elements.

$$V_c = \sum_{\substack{\text{All } i \\ \text{closed}}} \frac{\Delta h}{3} \left[A_i + A_{i+1} + \sqrt{A_i A_{i+1}} \right]$$

If a mound is represented by a set of closed contours the volume V_c is added to the open contour volume V_o, and if a depression is represented the volume V_c is subtracted from the open contour volume V_o.

Thus the volume bounded between the surface represented by open and closed contours and the reference plane is determined. If this is repeated for

25

before and after wear contour maps, using the same reference plane in both cases, the volume change due to wear is determined by subtraction of one volume from the other, provided the before and after wear contour information in the computer is to the correct scale and that any error in reorientation of the object after wear in the DIHC system is corrected. Scaling and orientation correction is achieved by having three dimensional reference points on the object of known two dimensional coordinates in the plane on which all three lie.

A calibration test of the system was carried out by applying it to a simple wear scar in an aluminium block. The volume change of the block due to the introduction of the wear scar according to the DIHC/computer method was within 4% of the volume change according to a weighting method.

The numerical methods so far described are applicable to simple surfaces. At present, more appropriate numerical methods for general surfaces are being developed where a surface is recorded in the form of a fine matrix of points, a height value being associated with each point. The height value of each point on the matrix is interpolated from the contour information, hence with scaling and orientation correction a method of imaging the difference between the before and after wear surface is being developed.

Acknowledgement

The authors are pleased to acknowledge the support of the Science Research Council for the work reported in this paper; and would like to thank Professor Swanson and Mr. K. Hardinge for providing the worn knee prostheses (Figs.3 and 4, and 5 and 6 respectively).

References

1. D. Dowson and B. Gillis (Leeds University) Private communication during 1978.

2. J.T. Scales and K.W.J. Wright. "Hip joint simulators used at Stanmore". One day meeting on joint simulation, Leeds (1976).

3. T. Tsuruta et al. Jap.J.Appl.Phys. 6, 661 (1976) and J.T. Atkinson and M.J. Lalor, 1st European congress on optics applied to metrology. Strasbourg SPIE Vol.136, (1977).

4. J.T. Atkinson, Ph.D. Thesis, Dept. M.M.P.E., Liverpool Polytechnic (1979).

5. J. Jaworzyn et al. S6. Gen.Serr.Int.ASS Dent Res, Washington D.C. April 1978.

6. J.T. Atkinson et al. For presentation at International Biomaterials Symposium, Clemson, South Carolina, April 28th, 1979.

7. A.A. Friesem, U.Levy and Y.Silberberg, "The Engineering Uses of Coherent Optics", E.R. Robertson (Ed.), C.U.P. (1976).

Quantitative Evaluation of Holographic Deformation Investigations in Experimental Orthopedics

U. Hanser

Orthopädische-Universitäts-Klinik, 6650 Homburg/Saar, Fed. Rep. of Germany

Deformation measurements by means of holographic interferometry are of special interest in the fields of osteosynthesis and endoprosthesis research (Fig. 1,2). For osteosynthesis with compression plates -the junction of bone fragments- holography gives immediately informations about the pattern deformation and thus, the conduction of forces in the system plate-screws fixed to the bone. The mechanical reaction of bone to the design of hip endoprosthesis stems under load can be made visible and in both cases quantitative information is given by calculating the lines of bending.

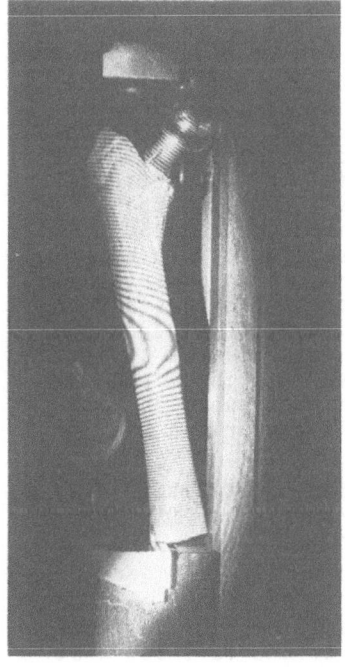

Fig. 1 Femur with hip endoprosthesis under axial load

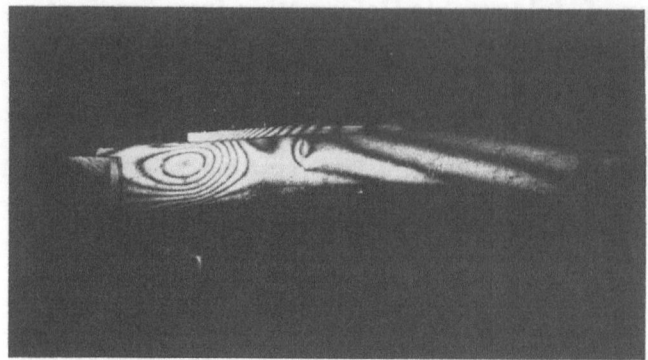

<u>Fig. 2</u> Tibia with osteotomy supplied with compression plate

In cases of alloarthroplastic replacement of hip joints or in
cases of bone fracture or osteotomy of bone, respectively,
treated with screws and compression plates, the functional bone
structure should not be influenced by endoprostheses or other
implants in order to avoid remodeling of bone caused by neutra-
lization of the supporting function. To rate the applicability
of implants we have to compare extension and stress distribution and thus the conduction of forces in supplied bone to
those in unsupplied bone. The acting forces are measured by
mechanical and electronical force transducers. For the determination of relative extension and upsetting of bone, the
strain gauges method is used since more than 30 years in experimental biomechanics. The determination of real stress distribution presents difficulties of experimental and theoretical
nature. The use of photoelasticity is possible only with homo-
genous bone models. For these reasons we indicated already in
1971 the applicability of holographic interferometry to determine the real stress distribution in human bone in experimental
orthopedics and we demonstrated our results in endoprosthesis
research. In the following year we showed double exposure holograms in osteosynthesis research. In 1976 we presented a film
demonstration about te determination of bending lines from femur supplied with endoprosthesis and tibia bone supplied with
autocompression plate, respectively, and in comparison that of
bending lines of unsupplied bone.

In the following we are showing examples for quantitative evaluations of typical double exposure holograms in the fields of
osteosynthesis and endoprosthesis research.

<u>Applications in osteosynthesis research</u>

Human tibia with osteotomy supplied with autocompression plate
(type MITTELMEIER) is deformed by driving of the left screw at
the end of the plate (Fig. 2). The lines of deformation are
determined from the number of the dark interference lines
(minima of interference), the wave length of the used laser
light (0.514 µm) and a geometrical factor. The quantitative
evaluations are given in Fig. 3 and 4.

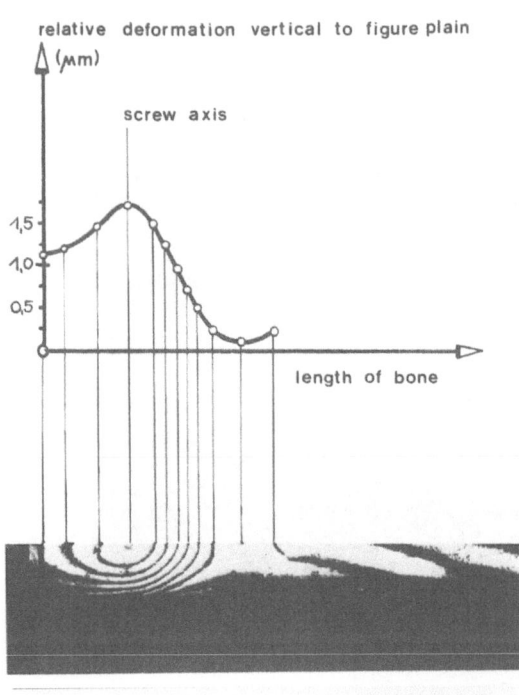

relative deformation vertical to figure plain

(μm)

screw axis

1,5
1,0
0,5

length of bone

Fig. 3
Relative deformation of tibia bone

Fig. 4
Relative deformation of tibia bone

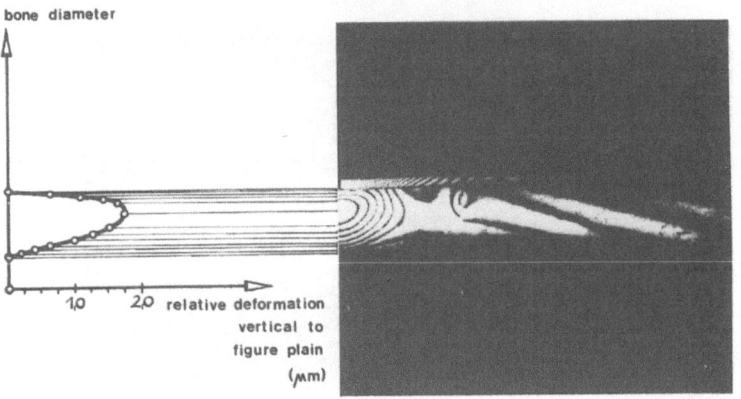

bone diameter

1,0   2,0   relative deformation vertical to figure plain (μm)

The next example shows the deformation of tibia without plate by bending on a 4-point-support.

The following figures 5, 6 and 7 are showing double exposure holograms with constant moment of bending. The lines of de-flexion are determined in analogical manner (Fig. 8).

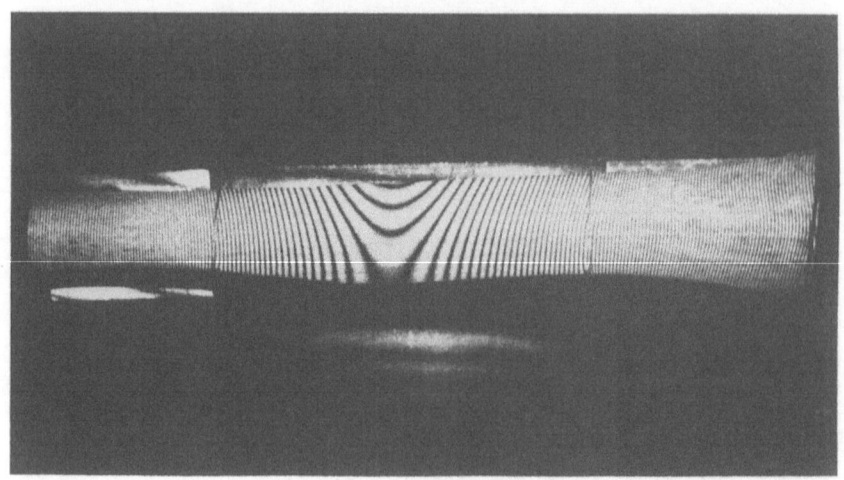

Fig. 5 Tibia unsupplied

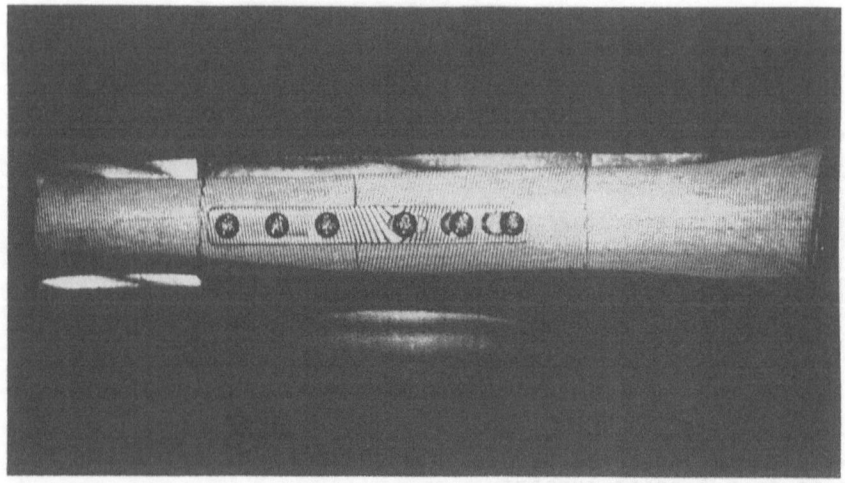

Fig. 6 Tibia with plate 1, thickness d 1

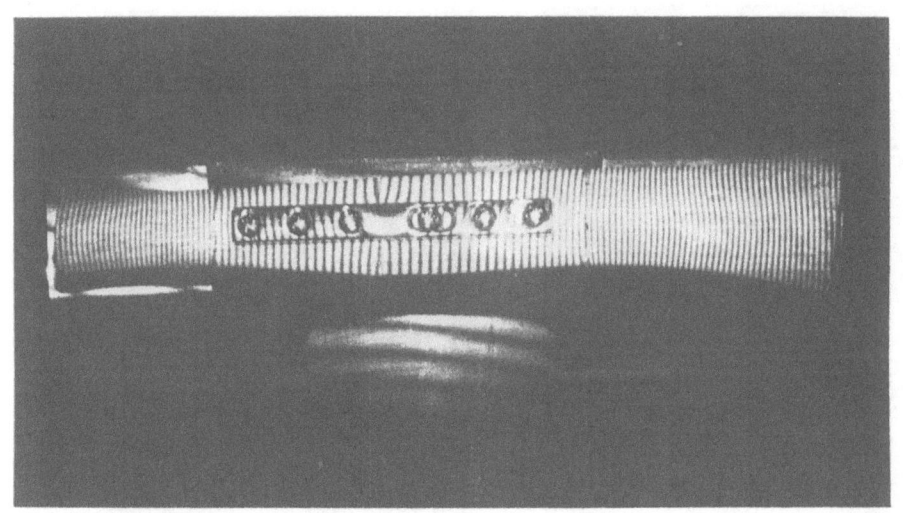

Fig. 7 Tibia with plate 2, thickness d 2

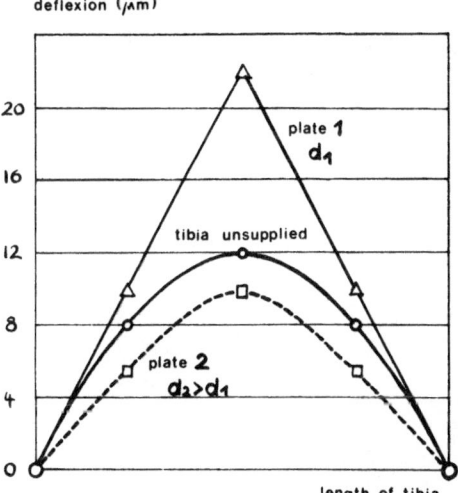

Fig. 8 From the comparison of the bending lines follows that the bending line of plate 2 approximates much better the ideal line of unsupplied tibia than the line of plate 1

Applications in endoprosthesis research

A human femur with and without hip endoprosthesis in a loading frame simulating the forces acting on the hip joint is deformed by bending. The bending lines are determined in analogical manner. The bending line of the unsupplied femur represents the ideal line for a hip prosthesis, which should alterate the functional structure of bone as little as possible. The double exposure holograms (Fig. 9 - 11) are evaluated as shown in Fig. 12. The femur gets more stiffness by application of endoprosthesis. The graphic representation of the deflection caused by bending shows that the bending line of prosthesis 2 approxi-

31

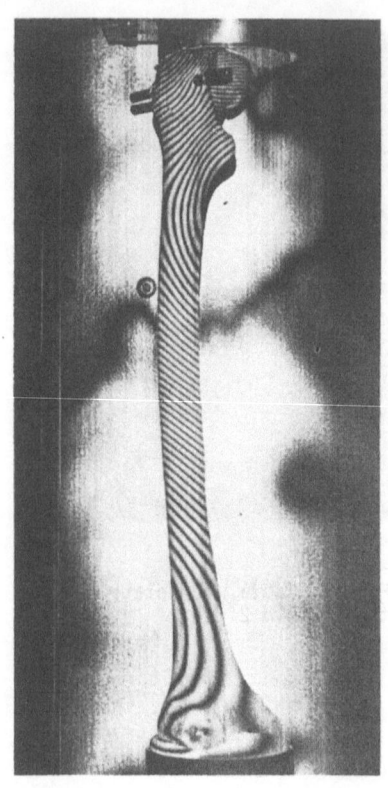

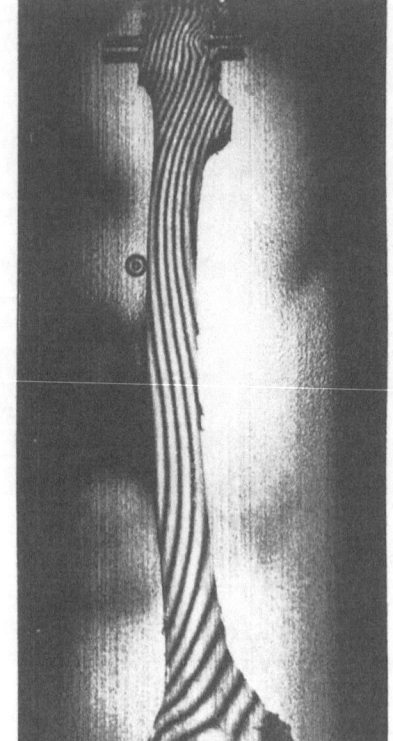

Fig. 9 unsupplied femur

Fig. 10 Femur with prosthesis 1

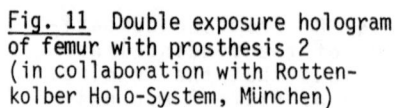

Fig. 11 Double exposure hologram of femur with prosthesis 2 (in collaboration with Rotten-kolber Holo-System, München)

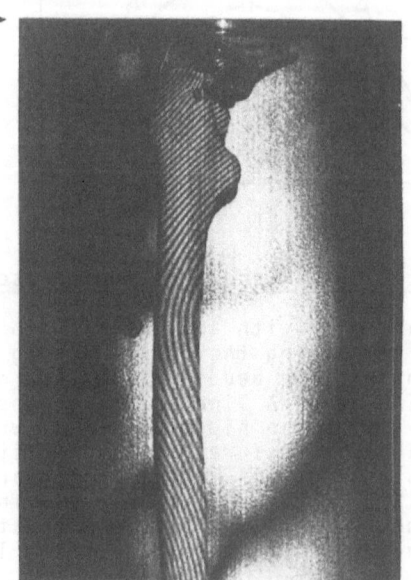

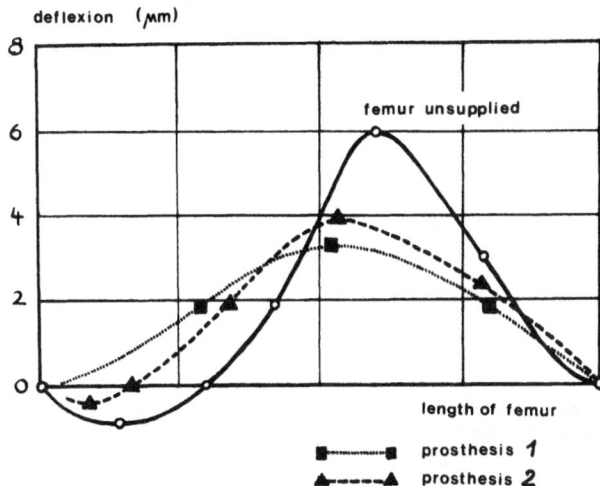

Fig. 12 Bending line of unsupplied femur and femur with
endoprosthesis 1 and 2

mates better the ideal line of the unsupplied femur than that
of prosthesis 1.

References

1. Hanser, U.: Spannungsoptische Untersuchungen bei der Osteo-
   synthese und Endoprothetik. Z. Orthop. 110 (1972) 871-876

2. Hanser, U., J. Harms u. H. Mittelmeier: Spannungsoptische
   und holographische Untersuchungen zur Biomechanik der
   Plattenosteosynthese. MOT 2 (1974) 47-50

3. Hanser, U.: Anwendung der holographischen Interferometrie
   in der experimentellen Orthopädie. Symp. 1976, SFB 88 Arbeitstagung Holographie in der Medizin, Münster Vol. 1
   (1977) 343-348

4. Hanser, U.: Lasertechnische Bestimmung von Verformungen in
   der experimentellen Biomechanik. Laser 77, Opto-Electronics
   Conference Proceedings, Guildford, Surrey IPC Sci.a.Techn.
   Press Ltd. (1977) 629-636

5. Hanser, U.: Holographische Bestimmung von Verformungen in
   der experimentellen Biomechanik. Biomedizinische Technik,
   Ergänzungsband Nr. 23 (1978) 186-187

6. Häusler, G., T. Schwenk, K. Seidel: Holografische Defor-
   mationsmessungen zur Optimierung von Hüftgelenkimplantaten.
   Symp. 1976, SFB 88 Arbeitstagung Holographie in der Medizin
   Münster, Vol. 1 (1977) 349-355

7. Steinbichler, H., Rottenkolber, H., Mönch, E.: Quantitative
   Auswertung von holographischen Interferogrammen. Laser +
   Elektro-Optik Nr. 5 (1973) 9-15

Holographic Investigation of Mechanical Characteristics of the Complex Leg-Foot in Conditions of Lesion and Reconstruction

D. Vukičević, V. Nikolić, S. Vukičević, J. Hančević, and Ž. Šućur

Institute of Physics, Department of Anatomy and Department of Surgery
University of Zagreb, Yu-41001 Zagreb, Šalata 11, Yugoslavia

1. Introduction

In our previous work we succeeded in showing the mechanical role of epiphyseal cartilage participating in the fundamental formation of the trajectorial structure which adapts itself to the activity of forces and stress distribution /1,2/.

Continuing our biomechanical investigations we have given special attention to the ankle joint. When analysing interferograms and sandwich holograms of the anatomical specimens of the axially loaded talocrural joints in plantar and dorsal flexions, we found that the talus in the plantar flexion was flattened with a saddle-shaped deformation. Under the same axial loading in the dorsal flexion, the talus moved downwards and backwards /2,3/. We also described the distribution of deformations at the distal end of the tibia, fibula and the entire ligamentary system, but since the axial loading was performed by directly loading the ankle joint specimen across the tibia and fibula, rather than indirectly across the tibia and the tibio-fibular joints we were not able to state that the results obtained reflected what actually happens in reality /3/.

The aim of this work is to describe stress distribution in the entire leg and foot and to evaluate the mechanical role of some supportive elements of this complex, as a verification of our previous findings.

2. Experimental Procedure

The subjects of the investigation were embalmed anatomical specimens of the leg and foot complex with removed muscles, but preserved terminal tendons and their attachments. Ligaments and interosseous tibio-fibular membrane were left intact. Their mechanical characteristics were not significantly affected, because they were kept in a wet microclimate. In few specimens we left a greater portion of soft tissues surrounding the bone (peroneal muscles, tendons of the posterior leg muscles, articular capsule of the talocrural joint) removing only the joint capsule between the deltoid and talofibular ligament in order to make the trochlea of talus visible. The tibial plateau was fixed by means of a dental acrylate to an aluminum plate. Simulated natural load was transfered to the leg through a steel ball by an hydraulic press.

At first we determined dislocations and deformations of the tibia, fibula, interosseous membrane and talus under conditions of progressively increased axial loading with varying pressure on the upper tibial articular surface ("O" model).

After recognizing the behaviour and supportive role of the investigated elements under axial loading the same specimens were looked at in conditions of lesion: drilling enlargement of the medullary cavity, osteotomy of the tibia and the fibula ("L" model/)

In the third stage, reconstructive interventions simulating a surgical procedure were performed: osteosynthesis by centro- medullary nailing and the AO plate ("R" model). Based on the evaluation of the changed map of interferometric fringes and in corresponding variation of stress-strain behaviour of the entire reconstructed system, compared with the intact model, we determined some parameters indicating the partially bearing capacities of the the tibia, fibula, interosseous membrane and talus.

Double exposure, real time and sandwich hologram interferometry were used. Taking into account optical properties of investigated specimens, their sizes (0.3-0.5 m) and considerable rigid body translations under the axial loading we have optimized the optical setup with the object being 36 cm far from the holoplate center (Fig. 1).

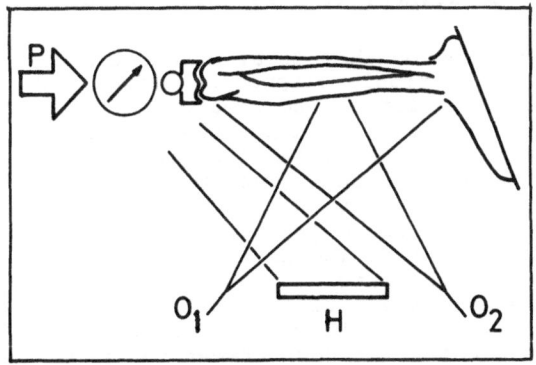

Fig. 1 Optical arrangement

Sensitivity for the rigid body translation in the axial direction was minimized. The specimens were placed in the central part of the holo diagram /4/, symmetrically around the normal onto the holo-plate center. Sensitivity factor variation around the average value of k=1.05 along the object surface, has been cut into half by two illuminating beams diverging from both sides of the sandwich plate holder /5/. In the data reduction procedure through static and dynamic analysis of interferograms, we took care of the average direction of illumination change for the distal and proximal parts of the leg-foot complex. Axial stress load magnitude needed for interferograms to be appropriately resolved was determined through the real time as well as sandwich hologram interferometry. Further on, we continued using the double exposure technique, because the main part of the rigid body motion was a simple translation and because of material saving reasons. In data reduction of double exposure interfero-

grams we used the BOONE and BACKER /6/-method of projecting the
reconstructed real picture onto the entrance of an simple ground
glass pin-hole camera and observing the interference pattern
which represents the point movement from the original to the
deformed position. Quantitative measures of the relative different movements of the supporting elements has been established
from the fringe density measurements /7/.

3. Experimental Results

3.1 Definition of the "O" Model-Intact Specimen

The tibia moved distally and pulled the talus, which moved medially, when the leg-foot complex was put under slight (2 kp) axial
loading (Fig. 2a). There were converging fringes on the medial
side of the upper third of the tibia which indicated bending towards the medial side, in the direction of the morphological curvature of the tibia. The bending occurred during greater axial
preloading (50-70 kp) because of the fixation and deformation of
the talus (Fig. 2c). The distal part of the tibia was more deformed than the proximal end and during greater axial loading (8 kp)
moved downwards and forwards in the direction of the head of the
talus. In the middle of the anterior margin of the tibial articular surface the tibia compressed the trochlea of the talus with
the small area, resulting in discontinued fringes which appeared
in that area like semi circles.

The fibula followed the direction of the tibial movement and
after the talus moved medially, the fibula slipped backwards and
laterally. The upper tibio-fibular joint showed greater mobility
than the lower joint. During all steps of loading, the fibula
was less dislocated and deformed than the tibia. The fibula showed no signs of bending, because of the great freedom of movement
in the upper tibio-fibular joint. Therefore, during small amounts
(2-3 kp) of axial load, the fibula remained passive and insignificantly participated in the load transfer. The entire function of
the fibula consisted of stretching the interosseous membrane which
also followed the downward movement of the tibia, pulling the fibula with it and thereby neutralizing the freedom of the fibula
in the upper joint. The membrane was stretched between the tibia
and fibula as a result of the dislocation and bending of the tibia, but because the fibula was passive the membrane slidd in
the direction of the leg's axis. It was stretched in the direction of its main fibre bundles at an angle of 20° to 30° running
towards the fibula distally beginning at the proximal end of the
tibia. The membrane and the fibula interrelate-the membrane stret-
ches the fibula and the fibula serves as a support for the func-
tioning of the membrane.

The zone of maximal shearing of the interosseous membrane was
located in the upper central part of the membrane, which was loaded the most (Fig. 2a). This particular zone divided the upper
portion of the membrane into the tibial and fibular area. There
were fewer fringes in the tibial area than in the fibular area,
again confirming the findings that the membrane near the fibula
was loaded the most and pulled the fibula downwards. The zone of
maximal shear began at the distal margin of the opening for the

anterior tibial artery near the fibula and ended at the small
opening of the membrane where the upper curvature of the tibia
finished. When the axial loading was increased the zone of maximal shear thickened (Fig. 2b), but its location did not change.
During maximal loadings the zone of shearing moved almost to the
margin of the fibula but slightly away from the interosseous margin of the tibia (Fig. 2c).

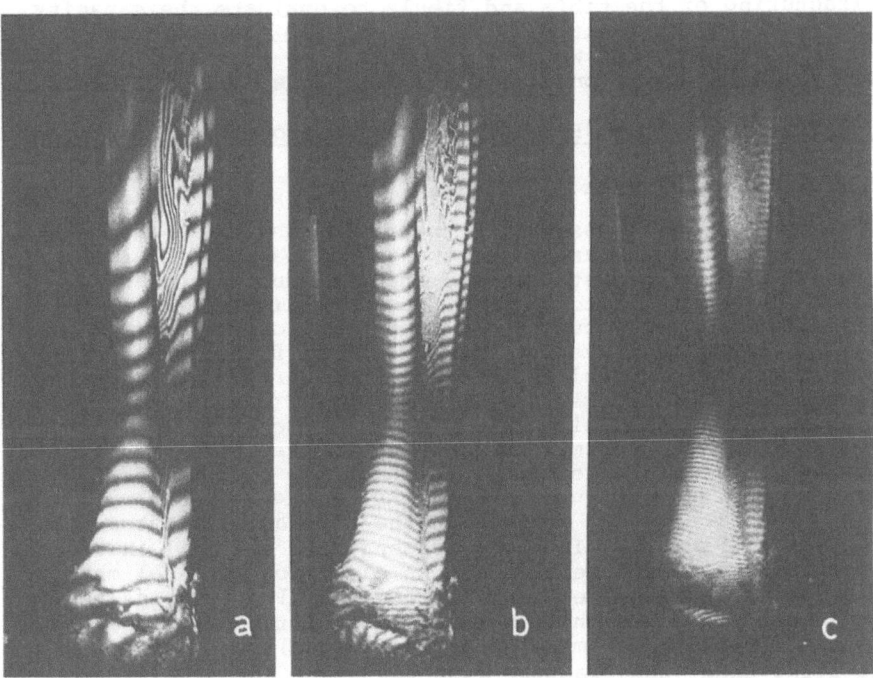

Fig. 2 The foot-leg complex under gradually increased loading.
The interosseous membrane is stretched between the tibia and fibula (a) pulling the passive fibula with it (b). The zone of shearing thickens with greater amount of loading (c).

The amount of preloading (0,30,50,70 kp) before the specimens
were progressively loaded did not influence the interacting behaviour and the type of deformation of individual parts of the
lower leg. Only the talus needed more preloading before it was
possible to observe the saddle-shaped deformation which occured
when the talus was fixed (moved downwards and medially).

Conclusion: The tibia bears almost all loading. The interosseous membrane is stretched by moving the tibia downwards, pulling
the passive fibula with it. The zone of shear is found in the upper part of the membrane which is stretched between the tibia and
fibula. The upper tibio-fibular joint shows greater mobility than
the lower one. Under the maximal load fibula is in average 30%
less deformed than tibia. Talus moves medially and downwards.

37

3.2   Definition of the "L" Model-Lesion of the Specimen

The analysis of the interferograms following drilling enlargement
of the medullary cavity of the tibia did not show any distinct
differences when compared with the "O" model. The way in which
the tibia was dislocated and deformed did not change, and the
behaviour of the fibula in load transference remained the same
in relation to the tibia and interosseous membrane. Simultaneously fracturing of the tibia and fibula to evaluate the capacity
of the membrane represented a problem because the dislocation of
the fractured bony ends was too large to give reproducible and
specific results, and consequently conclusions could not be drawn.

3.3   Definition of the "R" Model-Reconstruction of the Specimen

3.3.1   Interlocking nail

We thought that by approximately fracturing the tibia at the
junction of the middle and lower two-thirds part of the bone
(where fractures are most common), and introducing the interlocking nail according to Klemm we would obtain additional information about the behaviour of the fibula in load transference.
The interlocking nail was introduced and the fractured ends were
tightened as under normal operational procedures. One condition
of fracture was considered, making the cut at nominal 45° to the
long axis. The interlocking nail was fixed transversally with three
screws, one in the proximal end and two in the distal sections of
the tibia. Radiographs were used to check the position of the nail
and screws. At smaller loadings (2-3 kp) without high preloading
(30 kp), the density of fringes on the fibula was greater than
on the tibia (Fig. 3a). In the specimens with the nail in the distal part of the tibia, we observed an elevation (several microns)
in the region between the diaphysis and distal epiphysis (about
1 cm distally from the third screw). After introducing the third
screw the elevation was more intense (Fig. 3b). The part of the
tibia lying under the fracture line was given more loading and
angled towards the proximal fractured section. The upper part of
the tibia did not show any medial bending. The fibula was loaded
more than the tibia in the distal part, while the same dislocation was shown in the upper part as in the distal fractured part of
the tibia. Under smaller (4 kp) loadings the membrane was relaxed
in the upper part, while in the distal part it pulled the fibula
causing a greater dislocation and deformation of the lowered section of the tibia (Fig. 3a). The double amount of loading (8 kp)
caused increased saddle-shaped deformation of the talus. The earlier mentioned deformed zone between the diaphysis and distal epiphysis of the tibia became more distinct, with the diaphysis an-
gling towards the entire epiphysis. The proximal fracture end of
the tibia was significantly dislocated, but the two parts of bone were still not forced together. The fibula in the distal part
showed the same deformation found in the distal part of the tibia.
In the proximal direction, the density of fringes became sparser
and the gradual appearance of converging fringes indicated the
lateral bending of the upper part of the tibia. It is very interesting to note that only the part of fibula above the fractured
tibial line was bent, meaning that the location of the fracture

line influenced the length of the bent section of the fibular
diaphysis. The interferential fringes on the tibia and fibula below the fracture line moved interdependently in different directions, indicating the torque between the distal parts of the tibia
and fibula.

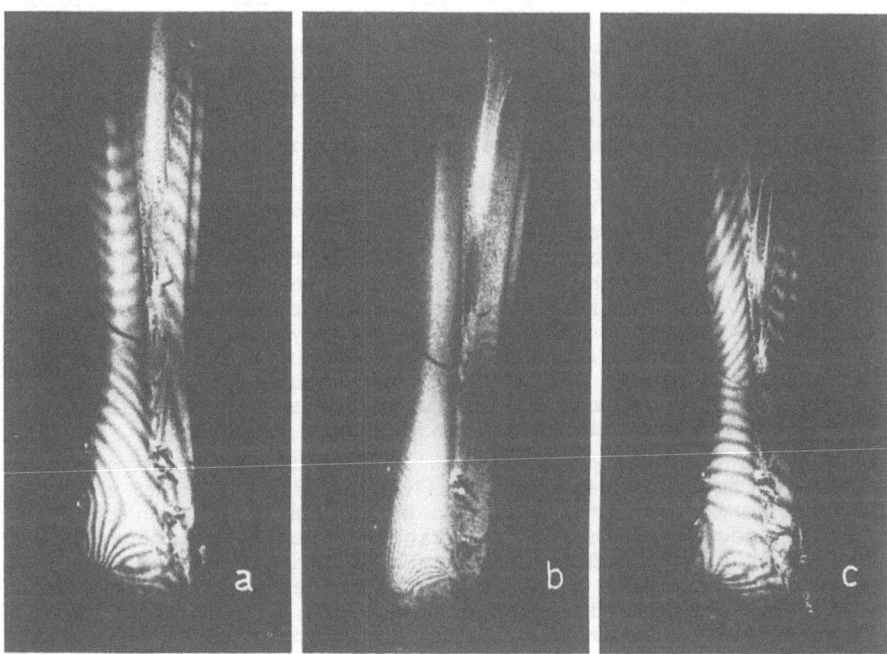

Fig. 3 Gradually loading of the specimen after reconstructing
of fractured tibial parts with an interlocking nail. The elevation under the third screw (a) becomes more distinct (b) by increased loading. Removing the screw, both fractured tibial ends
are forced together (c). The interosseous membrane is relaxed.

The interosseous membrane in the upper part did not show any
dislocation nor deformation since it was completely free. The talus showed complex fringes without the regular saddle-shaped deformation and dependent on the angulation of the fractured tibial part.

Conclusion: The fibula is more flexible than the tibia only
during minimal loading but as the loading increases the distal
parts of the fibula and tibia show identical dislocation and deformation. The fibula is again less deformed than the tibia. The
upper part of the interosseous membrane loses its function during
load transference, while the distal part helps the fibula to function and to pull the distal fractured part of the tibia. During
loading, with minimal preloading, the bending of the fibula depends on the location of the tibial fracture. In the lowered
fractured section there is a deformed zone between the diaphysis

and distal epiphysis which becomes more distinct by introducing the distal screw of the nail. Talus is not physiologically loaded.

In cases of increased preloading followed by a gradual increase of axial loading the entire fibula bent laterally while the fringes on the distal part of the fibular diaphysis below the fractured point went parallel with the axis of the fibula and were connected with corresponding fringes on the tibia. The angulated zone between the diaphysis and the distal epiphysis were more distinct. The concentration of fringes on the proximal part of the tibia was still less than on the distal part, meaning that the proximal screw still did not allow a good contact between the fractured ends. The interosseous membrane in the upper part continued to remain passive in the load transference.

Conclusion: Increased preloading does not significantly influence the interaction between individual parts of the lower leg following the fracture of the tibia and the amount of deformation changes insignificantly. It is obvious that the fibula is bended as a whole and there is an increase in deformation in the distal part of the membrane. The nail did not compensate the entire bearing capacity of the tibia as clearly-indicated by the behaviour of the fibula. The angulation of the diaphysis and the distal epiphysis of the tibia just like the angulation of the fractured tibial ends lead to irregular loading of the talus (saddle-shaped deformation with two distinct centers).

To confirm the role of the screw in relation to the interlocking nail, we removed the screw from the proximal part of the tibia and made a series of interferograms by gradually loading the anatomical specimens. In all investigated specimens independent of the amount of preloading, the distal epiphysis showed very dense fringes under the point of angulation already during minimal load. On both fractured tibial ends, the density of fringes was greater than that on the fibula. Interferential fringes did change the orientation on the fractured line, which then showed a good contact between both fragments in all parts of the diaphysial circumference (Fig. 3c). There were still zero order fringes on the membrane and of the more complex form. The deformation of the talus was again saddle-shaped. The angulation between the fragmented parts was still present but to a lesser extent.

Conclusion: When the upper screw is removed from the nail, the tibia relieves the fibula again, which under the same loading shows less deformation since the tibial bone fragments were forced together. The angulation of the bone fragments improves establishing the normal loading direction of the talus.

3.3.2 AO Plate

Finally we fixed the tibial fragments, as we did above, with the AO plate. Using the method of double exposure and sandwich hologram interferometry we found a greater concentration of fringes in the distal fractured part of the tibia (Fig. 4a). In the upper part of the proximal fractured end of the tibia, the fringes were dense only up to the first proximal screw of the plate. The density of the fringes became rarer towards the fracture line below this screw.

The density of the fringes rapidly declined just at the point of the first screw of the plate (Fig. 4b). The fringes of the distal fragment moved in opposite direction to those of the proximal fragment, indicating the angulation of both fragments. The fibular fringes were denser than those of both tibial parts and moved in the same direction as the fringes of the distal tibial section, but opposite to the fringes of the proximal fragment. This means that the entire fibula and the distal part of the tibia behaved as a unit. Under greater loadings, the talus showed the familiar saddle-shaped deformation, indicating it was correctly loaded.

Conclusion: The fibula is more loaded than the tibia, under all loading conditions. The plate only relieves part of the upper fragment just under the first screw, and does not allow good contact between the two parts, not even under maximal loadings. The fibula and the distal fragment behave like a unit with a somewhat larger deformation of the fibula. This means that the main loading capacity takes over the fibula and pulls the lower fractured part of the tibia with it.

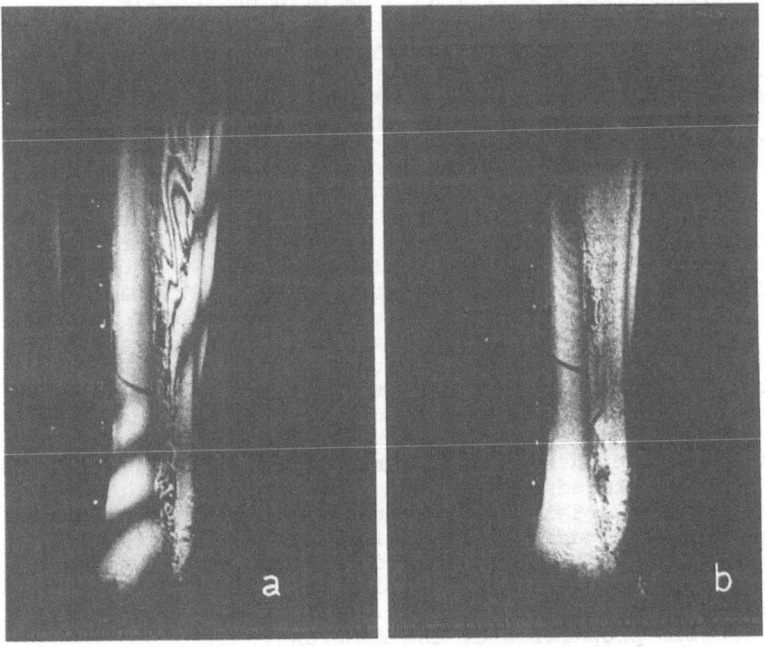

Fig. 4 The AO plate (a) relieves the upper fractured tibial part under the first screw (b). The fibula takes over the main loading capacity.

In the same anatomical specimens with insertion of plates following fibular fracturing, the relative deformation of the fragments did not change. Only the distal section was more deformed.

41

4. Discussion

If we consider the "O" model again, we see that the fibula does relieve the tibia although it behaves passively, within the deformation ratio range 10:6 to 10:7. This explains why with the moval of the fibula there is a significant thickening of the tibia, what was clearly shown by PEĆINA and MUFTIĆ /8/. The passivity of the fibula is a product of its anatomical relations and the increase mobility of the upper tibio-fibular joint. The mobility is to some extent reduced by the interosseous membrane, which provides resistance to the fibula moving upwards. The main role of the interosseous membrane is to serve as an auxiliary function of the fibula, clearly indicated by the adherence capacity of the membrane. The existence of the central shearing zone of the membrane can be explained by investigations of MINNS /9/ who found by studying the stress-strain characteristics of the tibio-fibular membrane, that it is 45 times stronger in the direction of its main bundles, than in the direction perpendicular to them, but much less extensible. In contrast to the findings of MINNS /10/, who has shown that on the surface of the tibial diaphysis in load transference there exists an area of high tensile stress corresponding to the tibial insertion of the interosseous membrane, our findings show exclusively the strain contentration in the upper tibial part of the membrane which is significantly less than the strain in the fibular part of the membrane. Both areas on the membrane are divided by the central zone of strain in which shearing is most obvious. By cutting the interosseous membrane KIMURA /11/ found that the tensile strain on the posterior side of the mid-shaft increased. This is clear, because our findings show that the role of the fibula in the load transference is enabled by the activity of the interosseous membrane. Thus, when the membrane is cut the tibia takes over a significant part of loading, which is usually transfered by the fibula. SARMIENTO et al. /12/ reported that the main function of the membrane is to hold the tibia and fibula together and when those bones are fractured, it contributes to the stability of the fracture, and helps to resist any displacement at the fracture site. It could be true, however, according to our findings, that this mechanism works only during the movement of fracture, because with the interruption of tibial continuity, the membrane does not show any signs of activity. It is relaxed by loss of tension of its main bundles in its entire upper two-thirds. However, its distal part (lower third) during maximal loading still maintains the functional relation between tibia and fibula. This data supports the mechanical function of the membrane and explains its role in load transference over the lower leg to the foot. This is not surprising knowing the tensile properties of the membrane.

WAGNER et al. /13/ by using the holographic interferometry method in an isolated specimen of the tibia and fibula with a partially preserved interosseous membrane (but fixed in plaster, without ankle joint) were able to demonstrate the fringes on the fibula and the membrane which support the tibia during axial loading. Their work shows only the possibility of holographic interferometry for illustrating the complex structures of the locomotor system. The way of load transmittance, however, was unnatural and the absence of the ankle joint does not allow us to draw any con-

clusions. They only reveal that the membrane is stretched between
the tibia and fibula, which can be expected from a simple force
consideration. Our results of investigating the entire anatomical
specimen, however, show that the activities on the membrane are
much more complex and cannot be anticipated only by knowing
their structural characteristics. Further more, this could indicate
that the use of holographic interferometry, although offering a
wide range of investigating possibilities, could lead to partial
and unreliable results, just as well as incorrect conclusions if
it is carried out under unfavorable and unnatural conditions.

Our findings show further that knowledge of the way in which
the talus is deformed, just as the direction of its dislocation in
changed conditions (introducing the nail or AO plate) could be
very important for evaluating the success of osteosynthesis. As
we have shown, inadequate osteosynthesis leads to completely
different deformation of the talus from the normal. This causes unnatural loading and could accelerate the damage of the articular
cartilage of the talocrural joint, as well as disturb the relations at the fractured line. Our previous findings agree with
our present results of talus deformation /2,3/, in conditions
of plantar and dorsal flexions of the ankle joints. As we have
demonstrated, under small axial load the talus moves first
medially and then downwards, and once fixed, it always shows a
saddle-shaped deformation. When the interlocking nail is inserted,
the talus is not physiologically loaded but there already exists
two areas of increased strain in the trochlear region. The AO
plate enables completely regular loading of the talus, while on
the other hand, it relieves the fracture line, especially on the
proximal fracture side of the tibia.

HARDING et al. /14/ using the method of double exposure showed that as compression between two parts of the tibia is increased, using a Mueller plate, there is greater loading at that part
of the fracture face nearest the plate and reduced loading away
from the plate. Therefore, its function is opposite to the
function of the AO plate which reduces loading under the first
screw and transfers it to the distal fragment.

That the knowledge of physiological loading is absolutely necessary for the analysis of different osteosynthetic devices has
been clearly shown. Likewise, the knowledge of ideal distribution
of interferential fringes under certain loading conditions (for
particular osteosynthetic devices as for certain surgical procedures using endoprosthesis) must be the starting point for the
evaluation, quality and success of the treatment in orthopedics,
as recommended by HANSER /15/ and HÄUSLER et al. /16/.

References

1. S.Vukičević, J.Hančević, D.Vukičević: Lij. vjes. 97, 16-21(1975)
2. D.Vukičević, J.Hančević, V.Nikolić, S.Vukičević: Proc. Electrococh. and Holography in Med., Münster (1976)
3. S.Vukičević, D.Vukičević, V.Nikolić, J.Hančević: Acta med.
   iug. 31, 251 (1977)
4. N.Abramson: Appl. Opt. 8, 1235 (1969)

5. N.Abramson: Appl. Opt. 16, 2521 (1977)
6. P.M.Boone, L.C.De Backer: Optic 37, 61 (1973)
7. E.B.Aleksandrov, A.M.Bonch-Bruevich: Sov. Phys.-Tech. Phys. 12, 258 (1967)
8. M.Pećina, O.Muftić: Proc. Ist Congr. Eur. Anthrop., Zagreb (1977)
9. R.J.Minns, J.A.A.Hunter: Acta orthop. scand. 47, 236 (1976)
10. R.J.Minns: Ph. D. Thesis, Dept Mechan. Eng., Liverpool (1975)
11. T.Kimura: J. Fac. Sci., Univ. Tokyo 5, 4, 319 (1974)
12. A.Sarmiento, L.Latta, A.Ziliolia, W.Sinclair: Clin. Orthop. 105, 309 (1974)
13. J.Wagner, J.Ebbeni, M.Clemens: Acta Orthop. Belg. 41, 24 (1975)
14. K.Hardinge, G.R.Bremble, M.J.Lalor: Proc. Electrococh. and Holography in Med., Münster (1976)
15. U.Hanser: Proc. Electrococh. and Holography in Med., Münster (1976)
16. G.Häusler, T.Schwenk, K.Seidel: proc. Electrococh. and Holography in Med., Münster (1976)

III. Moiré Topography

The Development and the Present Status of Moiré Topography

H. Takasaki

Research Institute of Electronics, Shizuoka University
3-5-1 Johoku, Hamamatsu 432, Japan

1. Introduction

The 1960s were the age of coherent optics. Many new optical techniques such as holography and speckle interferometry were realized by the use of coherent laser light. But it is also true that classical optics worked beautifully as well to solve some of the problems, coherent optical techniques were applied to. Moiré topography is one of the examples.

In 1969 TSURUTA [1] reported on a holographic contour generation. This technique is featured by recording Young's interference fringes, which are projected obliquely on an object, and then visualizing the contour lines of the object by interfering the diffracted wave of plus first with that of minus first order.

In a private discussion with the author, KASAHARA [2] pointed out that the contour fringes of TSURUTA's method was the moiré formed by the recorded interference fringes which were deformed following the shape of the object and interference fringes formed by the two collimated beams used in the reconstruction process. If this is the case, the same type of fringes must be visualized by merely placing a linear grating on an object and observing the shadow of the grating projected on the object through the grating.

This technique has been known in the field of stress analysis under the name of shadow moiré technique [3], and PIRODDA [4] had analyzed the conditions to obtain exact contour fringes of an object with high curvature.

In 1970 MEADOWS et al. and TAKASAKI published papers in the same journal, Applied Optics, in succession [5] [6], and the new era of modern moiré topography opened.

The characteristic difference between classical shadow moiré technique and modern moiré topography is in the subject to be measured. Shadow moiré technique has been used to measure solid, often almost flat, objects, whereas moiré topography is developed to measure live objects. One of the main problems in using moiré topography were found in the process to show contour lines of live objects by means of moiré fringes with good visibility.

Such problems were solved after all [7] [8], and the method spread rapidly in the medical field. For instance, a research group to study medical applications of moiré topography was formed in Japan in 1975, and a meeting is held regularly once a year.

In medical applications the objects are mostly live human bodies, which cause special problems. One of the requirements of the medical users of moiré topography is to have free space around the object under test. This is satisfied by moiré topography of the projection type [9] [10] [11]. Another method of the projection type, that has been proposed, utilizes the additive moiré effect [12].

A technique which is also regarded to be practical in medical use, because of its capability of posterior adjustment of the fringes, is moiré topography of the grating hologram type [13] [14]. In this technique the image of the grating projected on the object is photographed, and then an adjustable grating is interfered with the record to show up adjustable fringes.

The method of the grating hologram type is very similar to techniques which were published before the advent of modern moiré topography [15] [16]. The difference between these techniques is that the former uses incoherent light, whereas the latter uses coherent light. Thus, moiré topography has made a round trip and came back to its original point, but on a higher level.

Moiré topography is still developing. There are attempts to increase its capabilities by introducing TV techniques [17] [18] [19].

The process of development and practical know-how of moiré topography will be reported putting stress on medical applications, together with a bird's-eye view on various techniques.

2. Basic Form of Moiré Topography

2.1 Point Projection Type

An experimental set-up for the basic form of moiré topography is shown in Fig. 1.

A sculpture O is placed beyond a grating G, and is illuminated by a Xe-short-arc lamp L in the square housing in the center of the picture.

Moiré fringes which look like contour lines of the object are to be seen on the sculpture. The man standing at the right also sees moiré fringes on the sculpture. But the fringes observed by him and by us are different. Neither of them are the exact contour lines. Moiré fringes which show the exact contour lines are seen through a camera C which is placed beside the light source. The situation is schematically shown in Fig. 2

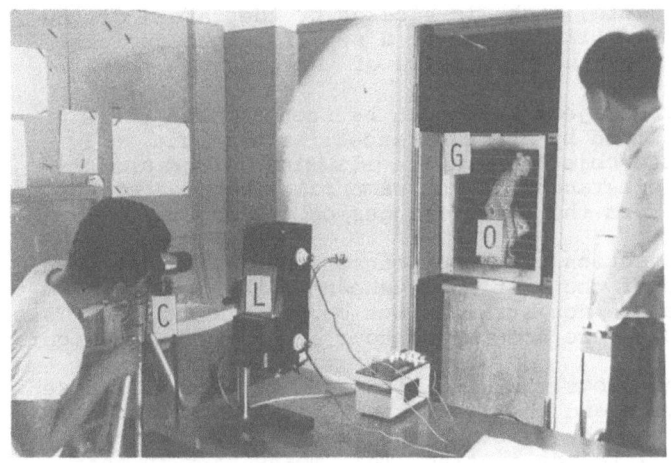

Fig. 1   Experimental set-up in its early stage

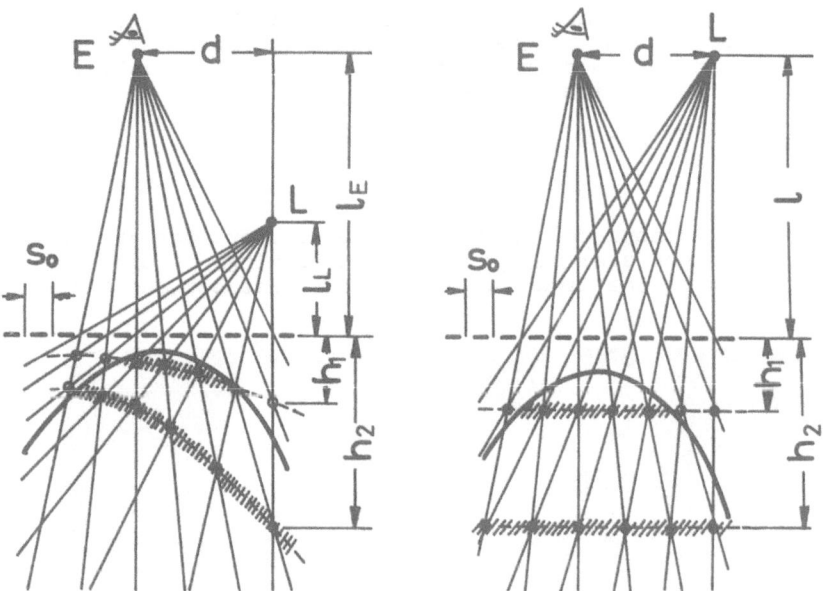

Fig. 2   Schematic diagram of the principle of the "basic form" of moiré topography

The left drawing shows the general situation where the light source L and the pupil E of the camera, or of the eye, are situated at points with different distances from the grating.

Since the illuminating light is blocked by the thread of the grating, the object is illuminated by a set of bright stripes of light coming through the clearances of the grating.

The regions on the object which can be seen through the grating are also limited by the clearances. As a result, only those portions of the object which are illuminated and are observed through the clearances at the same time are seen bright. These bright spots form the moiré fringes on the object.

The vertical projection of these points to the plane perpendicular to the thread of the grating is shown in the drawing by dotted lines. These are not straight but curved and fanning. This means that the fringes are not the contour lines of equal height.

The right drawing shows the situation in the particular case where the light source and the viewing points are positioned at the same distance from the grating. The dotted lines are now straight and parallel to the grating. This means that in this case the moiré fringes are contour lines with equal height of the object.

But they are not equally spaced. The height intervals between successive fringes increase with the distance from the grating. The exact distance from the grating to the N-th bright fringe is given by the simple geometrical formula (1), and the height difference between successive fringes by (2).

$$h_N = N l / (d / s_O - N) \tag{1}$$

$$h_{N+1} - h_N = (l\,d / s_O) / [(d/s_O - N - 1)(d/s_O\, N)] \tag{2}$$

From these equations we see that the key factors for the derivation of the height are: s_O, pitch of the grating; l, distance from the grating to the light source; d, distance between the

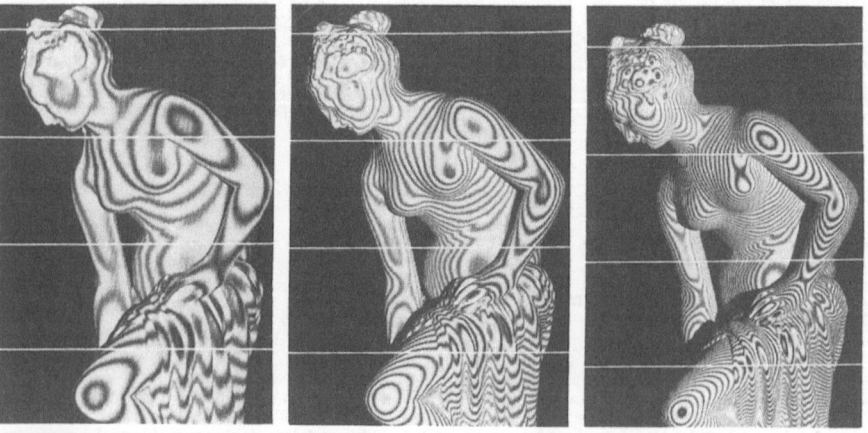

Fig. 3 Contour moiré fringes of different height difference adjusted by d

light source and the viewpoint measured vertically to the thread of the grating; and N, the order of the fringe under consideration.

It is also seen that the value of d/s_0 is an important parameter to characterize an instrument. This value is usually of the order of 1000, and so, the fringes are considered to be approximately contour lines of equal height difference, when the required accuracy is not too high.

The height interval between the fringes can easily be adjusted by changing d, as demonstrated by Fig. 3.

This basic form of moiré topography is known as the "Point Illumination Type" [5] [6] [7] [8]. Formation of the fringes of this type had been treated by PIRODDA [4], and a straightforward analysis is made by CHANG [20].

2.2 Moving Grating Technique

The pictures of contour moiré fringes thus obtained are deteriorated by two types of noises. The first is caused by the structure of the grating. This is demonstrated by the rough appearance of the wide fringes on the temple and left bust of the sculpture presented in the left picture of Fig. 3.

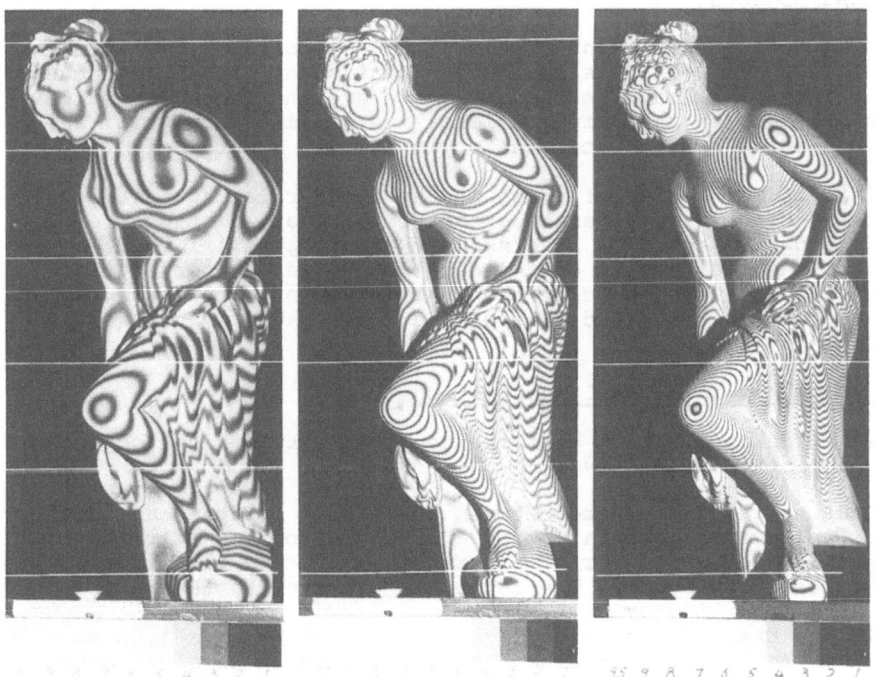

Fig. 4 Cancellation of noises by the "Moving Grating Technique"

The second is the moiré formed by the higher harmonic components of the grating and of the shadow projected on the object.
This is demonstrated by the queer patterns on the upper right
portion of chest and belly of the sculpture presented in the middle
and right picture of Fig. 3.

These noisy patterns move and change their shape following the
translation of the grating in its plane. So, these noises are all
cancelled by moving the grating during the exposure time of the
picture. Figure 4 demonstrates the results.

The blurred mark on the left foot in each picture shows that
the grating was moved by the length of the blur during the exposure.

This technique, which is now known under the name "Moving
Grating Technique", greatly improves the accuracy of measurement,
because the error in pitch of the grating is averaged over the
length of the movement.

2.3 Problems in Medical Applications

Moiré topography was introduced by the American Institute of
Physics [21], and a variety of practical uses has started since
then. But they were mostly industrial applications, because
further developments were needed to make the method practical for
medical objects.

In medical applications, the objects are mostly live human
bodies. The live skin is optically translucent, and this causes
the biggest problem. The shadow of the grating projected on live
skin diffuses on the skin and loses its grating structure. The
contour moiré fringes lose their visibility, accordingly.

Investigating the influence of the inter-relating parameters
several trials were given to preserve the grating structure of
the shadow projected on live skin at an acceptable level. The
process and the results of the development were reported elsewhere [7][8]. As a result, the application of moiré topography
has settled in the field of medicine, more rapidly than we expected.

2.4 Know-how in Medical Application

Know-how in application of moiré topography to medical objects
are summarized as follows:

Same height difference between successive fringes can be obtained either by combination of a fine grating and illumination
with small obliquity, or of a coarse grating and illumination
with greater obliquity. The latter is prefered to the former.

In the case of measuring a full size live body, a 1.50 mm
pitch is the best compromise. It is fine enough to give contour
moiré pictures of acceptable quality, and is coarse enough to

show fringes with good visibility on live skin over a wide range of 1 meter from the grating.

When the measurement is limited to rather flat parts of the body such as the back or the chest, a finer grating of 1.2 mm pitch can be used.

A grating is made either by stretching a solid thread or ruling it on a glass plate. The former is preferred to the latter for its low reflectance, and the latter is preferred to the former for its higher rigidity and stability.

The black spray "Nextel Velvet Coating 101-C 10 Black" of the 3M Co. is the best for cancelling the reflection of the grating made of a stretched thread.

It is advisable to supply the grating with a double-frame structure. The outer frame supports the tension of the thread, and the inner frame defines the flatness of the grating. Long mechanical threaded screws are conveniently used as bridges which define the pitch.

The choice of the diameter of the thread is an important factor. A thread diameter of exactly half of the pitch is the best choice.

The controlling factor of fringe visibility is the condition of the shadow of the grating projected on the object surface. Because the apparent width of the clear parts of the grating made of a stretched solid thread decreases with the angle of illumination, the width of illuminated stripes on the object surface is a little narrower than the width of dark shadow stripes. So, degradation of the grating structure of the shadow is prevented.

The light diffusion on live skin takes place inside the body. Since blood is red, blue light is more readily absorbed, and diffuses much less than red light. Hence, the use of blue light, or light of high color temperature such as that from a Xe arc lamp, helps to preserve good fringe visibility on live skin.

The light source for illumination is not necessarily a point, but can be a linear light source aligned with a straight line parallel to the thread of the grating. An illuminating system composed of two lamps positioned on the extreme ends of this straight line is preferred to prevent the light specularly reflected by the grating from getting into the observing system.

A grating with horizontal threads is preferred to one with vertical threads to measure live bodies in standing position, because the variation of direction and pitch of the shadow of a horizontal grating projected on a standing body is limited within much narrower range than that of a vertical grating. But the mechanical construction of the former system would be more complicated than that of the latter.

As stationary thread stretched close to the grating is helpful to determine the fringe order, which can be obtained from

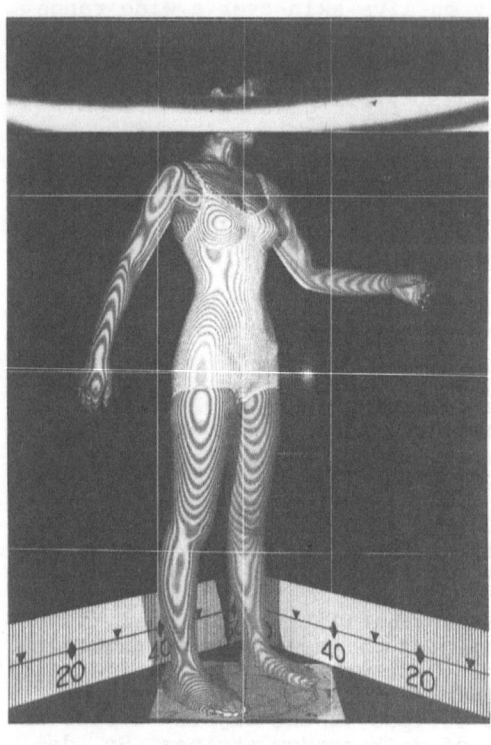

Fig. 5 Contour Moiré picture of a full size body

the apparent distance from the thread to the intersection points of the corresponding shadow and the fringes [7][8].

The exposure time should be less than 1/15 of a second. A camera of 6x6 cm^2 is ideal, but a smaller 35 mm camera is sufficient in most cases. A camera larger than 6x6 cm^2 is not recommended because of its smaller depth of focus.

An example of a contour moiré picture of a full size live body is shown in Fig. 5. The equipment was a prototype, and a grating with vertical threads was used. Technical data were l=160 cm, d=48 cm, s$_0$=1.5 mm, and diameter of the grating thread, actually a synthetic fishing line, 0.75 mm. The height difference between successive fringes is approximately 5.0 mm. The camera used was of a 6x6 cm^2 format, with a f=50 mm lens, which was stopped down to F 11. Exposure time was 1/8 of a second, and the film was a Kodak Tri X rated ASA 1600.

The central vertical line and the white squares were formed by stretching a white thread near to the grating surface. The shadow of the vertical thread can be seen on the upper chest of the model. This shadow line serves to determine the order of a fringe. The square was 40 by 40 cm, and serves to calibrate the magnification of a picture.

Penumbrae are seen on both sides of the lower legs. This means that there were two light sources on the left and right side of the camera. These were adjusted to give the same contour moiré fringes, and a shadow free picture [6].

The surface condition was natural. The model wore a corset of beige color, and was exposing her upper chest and arms. The fringes were retaining acceptable visibility on any portion of the body.

Further development of the instrument was made [8]. The new instrument is composed of two moiré cameras made of gratings with horizontal threads. The two moiré cameras are combined face to face with 60 cm clearance for a subject. A set of contour moiré pictures of a live body can be made from the front and from the back in less than 1/4 of a second which is just sufficient to immobilize the body.

Pairs of stereo pictures are also made with this instrument. Because the direction of the threads forming the grating is horizontal, the two cameras set in the same height have the same d, hence, the contour moiré fringes taken with the two cameras are the same, but the perspective is different, and form the stereo pair. A pair of stereo pictures of a live body taken with this instrument is shown in Fig. 6.

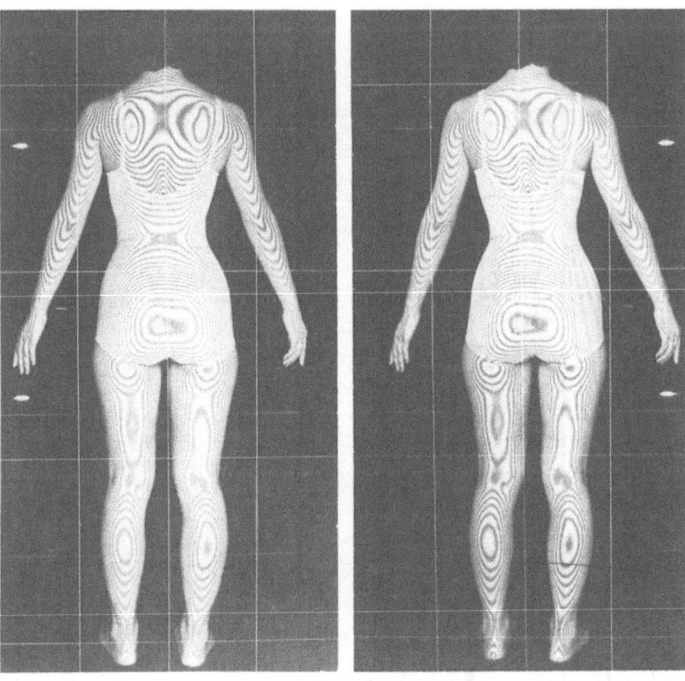

Fig. 6 A pair of stereo pictures of a live body

3. Projection Type of Moiré Topography

Moiré topography was for the first time applied to a live human body in order to evaluate the forming capacity of women's under wear. Since the subjects were healthy grown up persons, there was no difficulty in positioning them. But when the subjects were not healthy or not grown up, positioning the subjects in proper direction was a serious problem. Thus, it was desired to have free space around the object to be investigated.

This was satisfied by the "Projection Type"-technique, the principle of which is shown in Fig. 7. A small grating G_1 is projected obliquely on an object by a projection lens L_1. The projected grating is then imaged on a second grating in the image plane of the lens L_2, where the contour moiré is formed [9] [10] [11]. Large objects can be measured now with a small instrument having free space around the object. This type of instruments are commonly used, as well as those of the "Basic Form", in the medical field.

Another "Pojection Type"-method is proposed by HOVANESIAN [12]. Using this technique, the grating G_2 is projected on the object simultaneously with the grating G_1. Contour moiré fringes are observed directly on the object by virtue of nonlinear properties of the eye or the used photographic film.

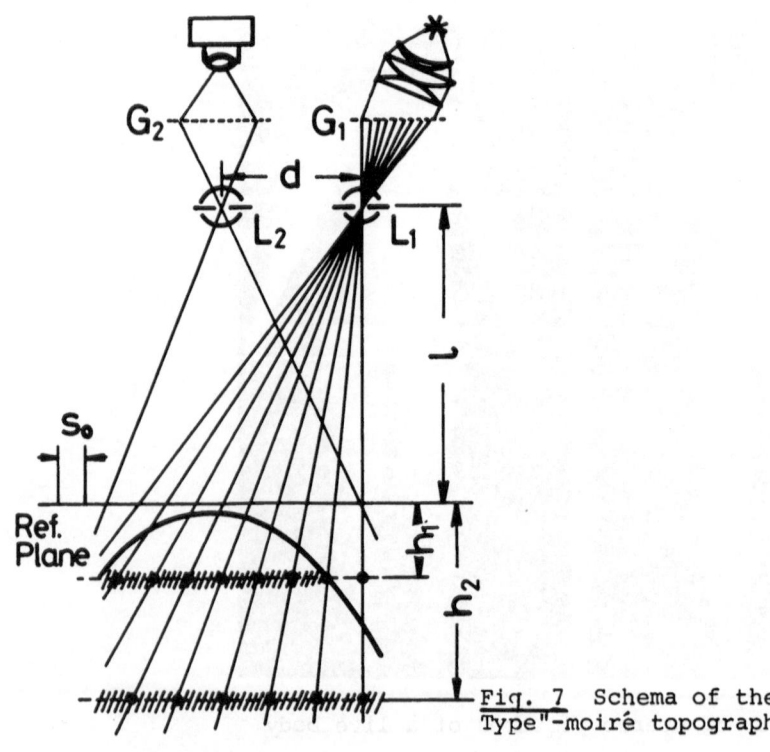

Fig. 7 Schema of the "Projection Type"-moiré topography

4. Grating Hologram Type of Moiré Topography

The projection type moiré topography has been used for early detection of scoliosis [22]. More than one hundred children have to be checked per hour in the field work. That means, there is not enough time for positioning each child. So, positioning is really a problem, especially when the children are less than twelve years old. Some twenty percent of the pictures of a recent field work were not squared at right angle to the bodies. Thus, a technique which makes it possible to adjust the contour fringes after recording is strongly wanted.

Moiré topography of the "Grating Hologram Type" solves this problem [13] [14] [15].

The linear grating which is projected and deformed following the shape of the object is photographed as shown in the left picture of Fig. 8. Informations on the body shape are stored in the picture in the curvature and spacing of the recorded grating at each point of the body. Hence, the photographic record is named "Grating Hologram".

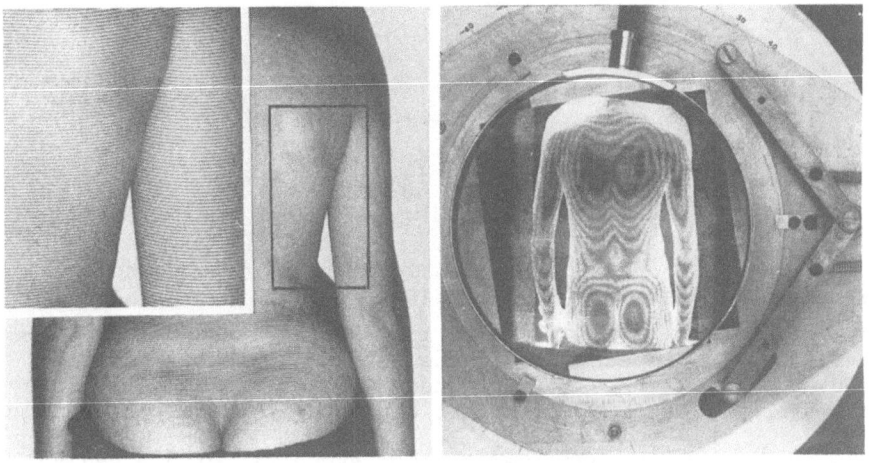

Fig. 8 Grating hologram of a live body (left); viewer (right)

The grating hologram is then interfered with an adjustable grating, which is made by overlapping two identical linear gratings, and changing the relative angle between them. The surface of one of the gratings is roughened with # 600 emergy powder to make the composite grating light-scattering. The grating hologram is projected on the composite grating with a projector, and the contour fringes are visualized as shown in the right picture of Fig. 8.

The adjustability of the fringes is demonstrated in Fig. 9. The left and right pictures of the middle row show fringes recorded in the cases the patient was facing towards left or right.

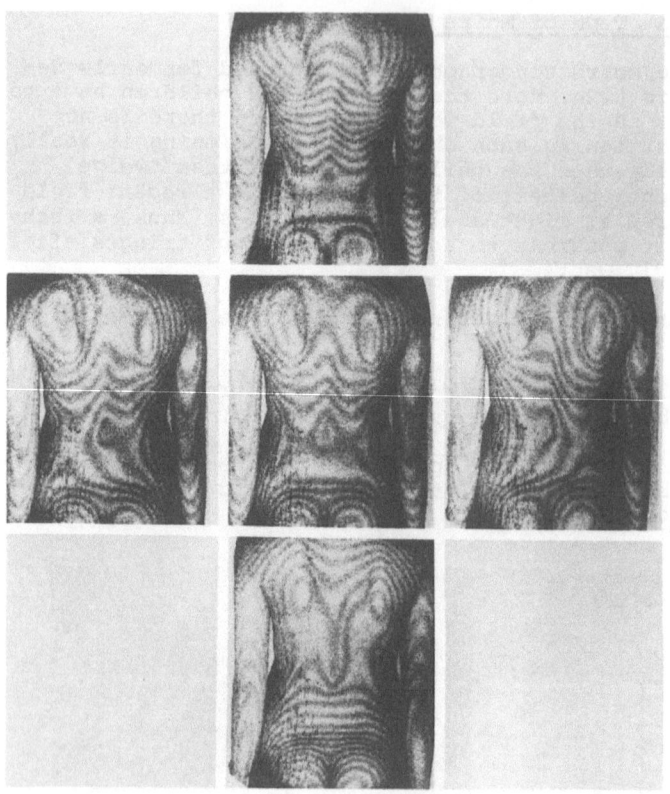

Fig. 9 Contour moiré pictures obtained from a grating hologram
at different adjustments of the viewer

Such fringes can appear also if the subject's back is normal-
shaped. So, it is impossible to judge whether the subject has a
normal shape of the body or not at a glance at these pictures.

The fringes can be adjusted to show symmetric circles on the
hips of the subject, as shown in the picture in the center. The
fringes at the *scapula* are now unsymmetrical showing the ano-
malous shape of the body. The anomaly is further scrutinized by
changing the slant angle cutting the body as shown in the pictures
in the center line. This slant adjustment is made by changing the
pitch of the composite adjustable grating, whereas the left to
right rotation adjustment is made by changing the angle of the
grating.

Technical data are summerized as follows. The pitch of the
grating to be projected is 0.125 mm. The projection distance
is 2.5 m, and the projection lens is a wide-angle f=150 mm lens.
A 35 mm camera with a f=135 mm lens is used for recording the
grating hologram. The grating is projected with its threads in
horizontal direction. The projector is put on the floor below the

camera which is set independently on the ground. This configuration is adopted to avoid vibration, but an inverted configuration is preferred from the point of picture quality because the latter gives more illumination on the shoulders which are important parts for the diagnosis. The shutter speed is 1/60 of a second using a photographic film rated ASA 3200. The lens opening is F 6.3 for projection and F 5.6 for recording. A 750 W iodine lamp is used for projection. The adjustable grating is made using glass gratings of 12 line pairs of equal width[1], and the grating hologram is projected on the composite grating using a low distortion macro lens.

5. Bird's-Eye View of the Development of Moiré Topography

The process of the development of various techniques of moiré topography is graphically shown in Fig. 10.

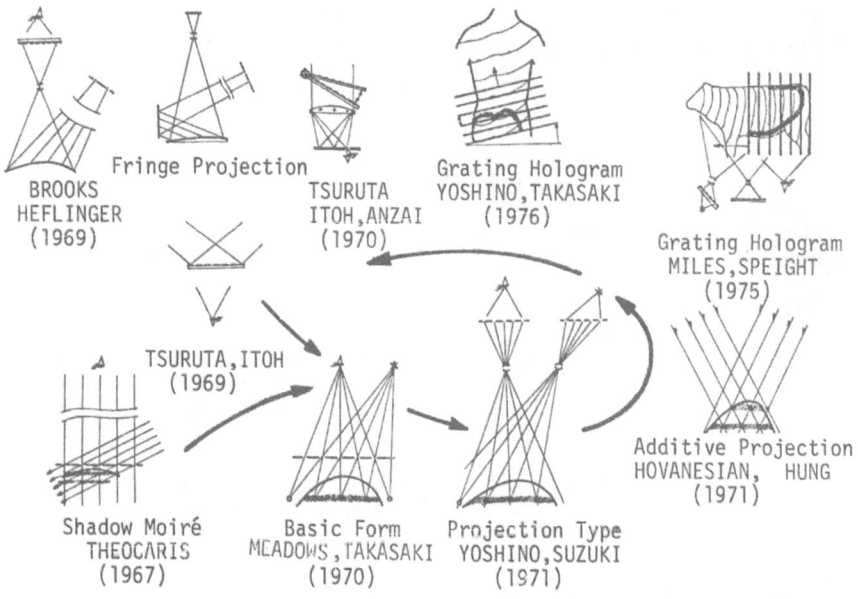

Fig. 10 Schematic drawing of the development of moiré topography

Starting from the "Basic Form", the "Projection Type" was developed to have free space around the object, and to adjust the fringes. Both of the techniques visualize the fringes in real-time.

The real-time character was stressed as a merit at the time of advent of these techniques. But this can be a demerit as well, because the contour fringes which are observed at the moment when a picture is shot are frozen, and there is no way to adjust the fringes after the record is made. The "Grating Hologram Type" was

---

1 Edmund Scientific Co. Barrington, N.J, U.S.A.

developed to acquire "posterior adjustability" but at the sacrifice of the real-time feature.

It seems that more and more freedom and capabilities are attained with the development. But it must be remembered that more and more of the merits of the "Basic Form" are traded away in order to achieve the new capabilities.

The greatest improvement was made by the development from the "Basic Form" to the "Projection Type". In the "Basic Form", the fringes are formed by the ray optics based on the relative relation of the grating and its shadow. The camera participates in the measurement only in taking pictures of the fringes already formed, and does not affect the measurement as far as the height of the object is concerned. An accuracy of 0.2 percent is proved using this method.

In the "Projection Type", the acuity and distortion of the lenses used play an essential role in the measurement. The grating must be projected clearly on the object, and then imaged again with good visibility. This state of acuity is never required for the lens of the camera in the "Basic Form".

The "Grating Hologram Type" involves even more optical elements than the "Projection Type", and imperfection of each element affects the result. It is not practical to try to obtain an accuracy better than 1-2 percent with the "Projection Type" or the "Grating Hologram Type", because it requires optical elements of high quality and highly accurate and stable construction.

The choice of a technique should be based on the clear understanding of what is needed and what is not needed. When the posterior adjustability of the fringes is strongly wanted, and the requirements of quantitative accuracy are not too stringent, the "Grating Hologram Type" can be the answer. If high accuracy is wanted, the answer would be the "Basic Form".

Good mutual understanding and collaboration among those people who need and those who develop a technique is most important to transform an idea into a practical technique.

6. Modern Trends in Moiré Topography

There is a new movement in moiré topography, that is the introduction of TV techniques [17] [18] [19]. IDESAWA and others applied computer processing techniques to the grating hologram after "reading" it with a low distortion image device developed by them. They also analyzed the three dimensional position of the fringes to be obtained by general combination of projection and viewing systems [17].

There are possibilities of on-line data processing such as calculating intersection profiles, the surface area, and the volume of live objects. The identification of concave and convex curvature of the surface is also possible by means of moiré topography [23].

58

Solid state image devices made of a square array of small photo-sensitive elements are promising for use in such "Video-Moiré Topography", because of their inherent dimensional stability and distortion free character. The success of "Video-Moiré Topography" depends upon how far the new capabilities, which cannot be free from some problems such as cost and accuracy, can be a-dapted to the needs of the users.

References

1. T. Tsuruta: Opt. Commun. 1, 34 (1969)
2. T.Kasahara: Private discussion with the author (1969)
3. P.S. Theocaris: Experimental Mechanics 7, 289 (1967)
4. L. Pirodda: Revista di Ingegneria 12, 913 (1968)
5. D.M. Meadows, W.O. Johnson, J.B. Allen: Appl. Opt. 9, 942 (1970)
6. H. Takasaki: Appl. Opt. 9, 1467 (1970)
7. H. Takasaki: Appl. Opt. 12, 845 (1973)
8. H. Takasaki: J. Amer. Soc. Photogrammetry XLI, 1527 (1975)
9. Y. Yoshino: Kogaku 1, 277 (1972) in Japanese
10. P. Benoit, E. Mathiew, J. Hormiere, A. Thomas: Nouv. Rev. Opt. 6, 67 (1975)
11. M. Suzuki, K. Suzuki: Koseido 3, 1 (1972) in Japanese
12. J.D. Hovanesian, Y.Y. Hung: Appl. Opt. 10, 2734 (1971)
13. C.A. Miles, B.S. Spieght: J. Phys. E 8, 773 (1975)
14. Y. Yoshino, M. Tsukiji, H. Takasaki: Appl. Opt. 15, 2414 (1976)
15. R.E. Brooks, L.O. Heflinger: Appl. Opt. 8, 935 (1969)
16. T. Tsuruta, Y. Itoh, S. Anzai: Appl. Opt. 9, 2802 (1970)
17. M. Idesawa, T. Yatagai, T. Soma: Appl. Opt. 16, 2152 (1977)
18. T. Ueda: Trans. Electronics and Communication Eng. Japan J 61 D, 299 (1978)
19. H. Kugel, F. Lanzl: this issue
20. C. Chang: Appl. Opt. 14, 177 (1975)
21. News from the American Institute of Physics, 31 July 1970
22. S. Inoue, H. Tsuji, Y. Otsuka, H. Suzuki, A. Shinoto: Seikei-geka 28, 746 (1977)
23. G. Windischbauer: this issue

Video-Electronic Generation of Real Time Moiré Topograms

H. Kugel and F. Lanzl

Institut für Angewandte Physik, Universität Hamburg
Jungiusstr. 11, D-2000 Hamburg 36, Fed. Rep. of Germany

Abstract

A system generating Moiré topograms in real time is described,
which delivers contour maps of the human back for medical applications. The generating gratings are eliminated and the Moiré
fringes are isolated to allow computerized evaluation of the contour map. The sign of the contour lines is determined to enable
distinction between elevation and depression.

Experimental Arrangement

The arrangement of the optical elements is based on the apparatus
for diagnosis and documentation of deformations of the human spine
which has been described by DRERUP [1]. The Moiré fringes are generated as proposed by TAKASAKI [2] with some alterations [3].

The set-up described here consists of a projection system casting a grating onto the surface under test, a video camera acting
as observer and an electronic system which generates the reference
grating electronically, produces the Moiré fringes by multiplication of the two gratings, eliminates the gratings from the contour
map and determines the sign of the contour lines. The result is
displayed on a monitor, but direct evaluation of the obtained data
in a computer is possible as well.

In the diagram of the optical arrangement (Fig.1) l is the distance between the nodal points O of the observation system and P
of the projection system, h is the distance between OP and an arbitrary virtual plane parallel to OP, and p is the period of the
grating cast onto this plane by the projector. The variable z is
the depth of the observed point on the object behind this plane.

The reference grating of appropriate pitch and direction is
simulated by a grating generator, consisting of a section deli-
vering switching signals and a video switch that writes black every second TV line. The reference grating thus is parallel to the
lines. Utilisation of TV systems is described in [4] and [5].

Generation and Isolation of Moiré Fringes

Observing a figure in the set-up, a camera receives an intensity
signal - converted into a voltage signal- (PG indicates projection
grating)

$$PG\pm = I(x)(\frac{1}{2} \pm \frac{1}{2}\cos\frac{2\pi}{p}(x - \frac{1 \cdot z}{h+z})) + I_U(x) \tag{1}$$

along a direction x perpendicular to the grating lines. I(x) is the intensity of the object (the y-dependence is omitted), the sum in brackets is the modulation caused by the grating cast onto the figure, which contains information about the interesting variable z in the phase of the cosine, and $I_U(x)$ is an intensity underground lying under the black portions of the grating, caused by optical and electronical distortion effects (Fig.2).

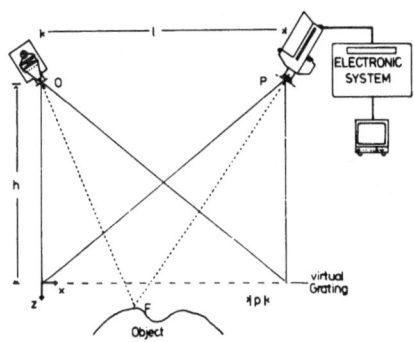

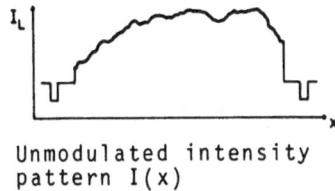

Unmodulated intensity pattern I(x)

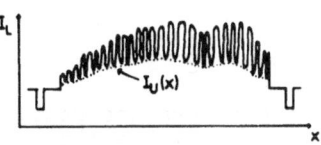

Fig.1 Optical arrangement

Fig.2 Intensity pattern according (1)

The grating delivered by the grating generator has approximately the shape (EG indicates electronic grating)

$$EG\pm = \frac{1}{2} \pm \frac{1}{2}\cos\frac{2\pi}{p}x \ . \tag{2}$$

The amplification of the video switch is set to unity. The exchange of + and - corresponds to a phase shift of 180^0.

The processing of the camera signal (1) by the video switch is equivalent to a multiplication of the signal with the grating (2). After the substitutions

$$\frac{2\pi}{p} = \gamma \ ; \quad - \frac{2\pi 1 z}{ph+pz} = \delta(z) \tag{3}$$

the multiplication of PG+ with EG+ results in

$$PG+ \cdot EG+ = \frac{I(x)}{4} + \frac{I_U(x)}{2} \qquad \text{(object + underground)} \tag{4}$$

$$+ (\frac{I(x)}{4} + \frac{I_U(x)}{2})\cos\gamma x \qquad \text{(reference grating)}$$

61

$$+ \frac{I(x)}{4}\cos(\gamma x + \delta(z)) \qquad \text{(projected grating)}$$

$$+ \frac{I(x)}{8}\cos(2\gamma x + \delta(z)) \qquad \text{(grating with double frequency)}$$

$$+ \frac{I(x)}{8}\cos\delta(z) \qquad \text{(Moiré fringe)} \qquad [6].$$

To obtain a clearly detectable contour map, it is necessary to eliminate all terms containing the grating signal. To allow computerized evaluation of the contour map, the object and the underground intensity must be eliminated as well. This is done by a combination of projected and electronic gratings according to

$$(PG+ \cdot EG+ + PG- \cdot EG-) - (PG+ \cdot EG- + PG- \cdot EG+) \qquad (5)$$

Each bracket represents a fringe pattern with eliminated reference and projected grating. The difference of these fringe patterns results in

$$I_{BIAS} + \frac{I(x)}{2}\cos(2\gamma x + \delta(z)) + \frac{I(x)}{2}\cos\delta(z) \qquad (6)$$

The second term has double frequency of the generating gratings, i.e. this term has the same period as the TV lines and is not resolved by the TV system. So the result of this operation is the pure Moiré pattern with an electronically introduced bias to prevent negative intensities. The results are shown in Fig.7.

Realisation

To obtain the terms in the brackets it is necessary to have PG+, PG-, EG+, and EG- available simultaneously. The change of the sign of the cosine in the gratings is equivalent to a phase shift of 180° in x-direction.

EG+ and EG- are easily obtained at the output and the inverse output of the grating generator.

To obtain PG+ and PG- simultaneously it is necessary for the observing system to record both the projection grating and the shifted grating at the same time without changing the observed intensity of illumination.

This problem is solved using a colour TV system. A colourless grating with black lines is projected with blue light. The blue channel of the camera records the intensity PG+' (ignoring the underground $I_u(x)$)

$$PG+' = I^{blue}(x)(\frac{1}{2} + \frac{1}{2}\cos(\gamma x + \delta(z)) . \qquad (7)$$

Using the same light source and the same objective the figure is illuminated with red light in an optically split path without grating (Fig.3).

The sensitivity of the red channel is adjusted so that it records

$$I^{red}(x) = I^{blue}(x) = I(x) \ . \tag{8}$$

In the camera control unit the signal of the blue channel is subtracted from the signal in the red channel, yielding

$$PG-' = I(x)(\tfrac{1}{2} - \tfrac{1}{2}\cos(\gamma x + \delta(z))). \tag{9}$$

By appropriate adjustment of the sensitivities and amplifications in the camera control unit the intensity underground of the difference signal

$$I_U^{red-blue}(x) = I_U^{blue}(x) = I_U(x) \tag{10}$$

is obtained. So four signals are available.

$$PG+ = I(x)(\tfrac{1}{2} + \tfrac{1}{2}\cos(\gamma x + \delta(z)))+ I_U(x) \tag{11}$$

$$PG- = I(x)(\tfrac{1}{2} - \tfrac{1}{2}\cos(\gamma x + \delta(z)))+ I_U(x)$$

$$EG+ = \qquad \tfrac{1}{2} + \tfrac{1}{2}\cos\gamma x$$

$$EG- = \qquad \tfrac{1}{2} - \tfrac{1}{2}\cos\gamma x$$

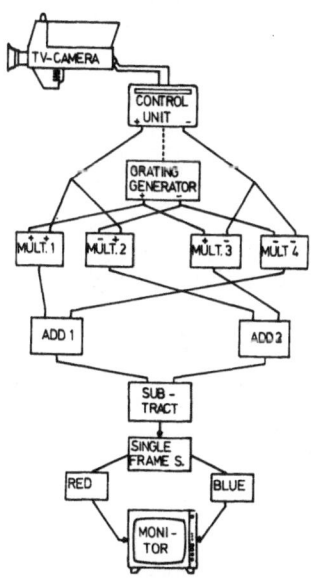

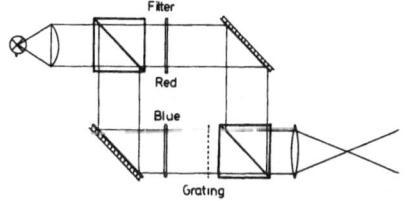

Fig.3 Projection system

Fig.4 Electronical set-up

Multiplication, addition and subtraction is done simultaneously by the electronic equipment as indicated in Fig.4.

The used TV system is a Philips LDH 20, the projected grid is the photograph of a raster foil on a glass plate. The arrangement with the data l = 190 cm, h = 260 cm, p = 3,4 cm, f(camera)= 50mm, f(project.)=60 mm delivers a distance in depth of about 4 mm between two consecutive Moiré fringes.

Sign Determination

The sign of the contour lines is determined taking advantage of a special effect of standard TV. Each image of 512 lines consists of two consecutively displayed images of 256 lines, called single frames. Each line of one single frame is placed in the space between two lines of the other single frame (Fig.5).

For our purpose 128 lines of each single frame, i.e. every second line, is written black. In every second single frame the grating lines are shifted half a line to the right - as the TV screen is installed so that the lines are vertical - compared with the first single frame, equivalent to a phase shift of 90°.

Shifting only one grating in an arrangement as the one described here corresponds to shifting the Moire fringes along the z-axis. The contour lines in the second single frame lie a quarter of a period further down in z-direction than in the first single frame.

Encoding the single frames with different colours enables the observer to distinguish between depression and elevation when observing the image on a colour monitor. The Moiré fringes are two-coloured in this case, with one colour always pointing upwards, the other colour pointing downwards (Fig.6).

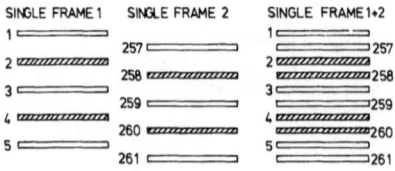

Fig.5 Location of TV lines

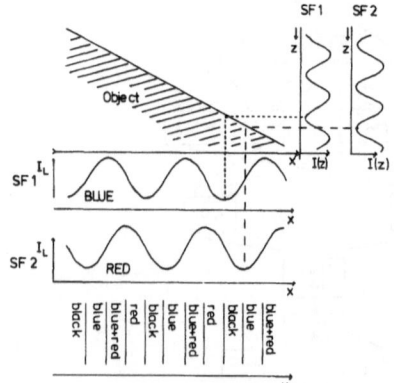

Fig.6 Determination of elevation and depression

Ref.[7] to [10] describe other methods of sign determination.

Results

The original and resulting images displayed on the TV screen are
shown in Fig.7. The measured object is a plaster cast of a patient suffering from scoliosis.

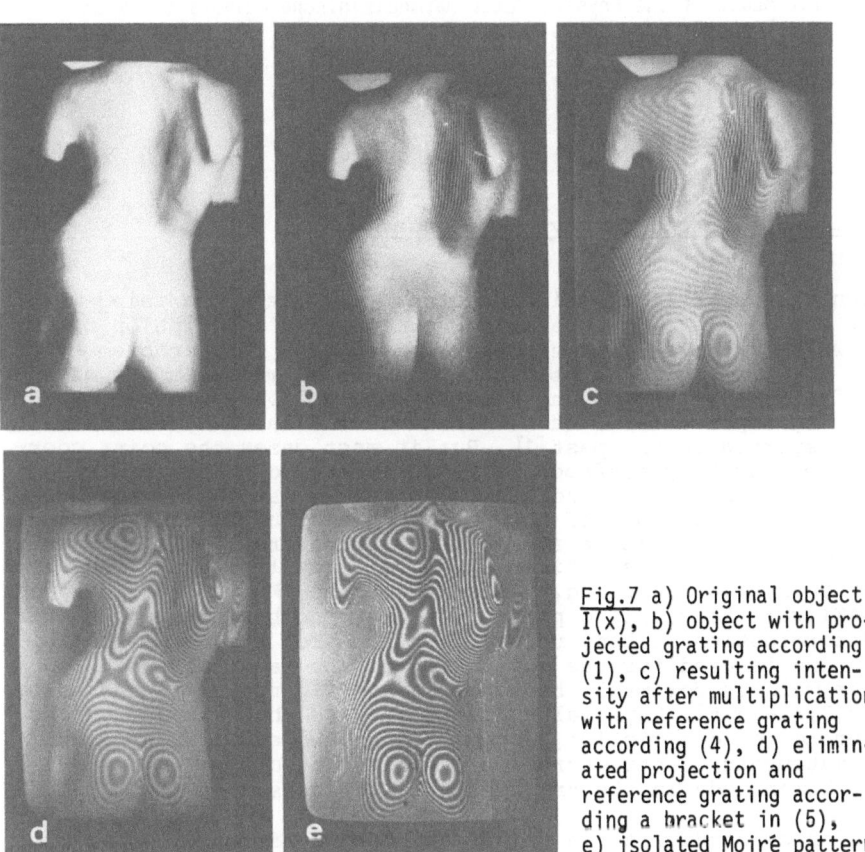

Fig.7 a) Original object
I(x), b) object with pro-
jected grating according
(1), c) resulting inten-
sity after multiplication
with reference grating
according (4), d) elimin-
ated projection and
reference grating accor-
ding a bracket in (5),
e) isolated Moiré pattern
according (6)

References

1. H. Takasaki: Appl. Opt. *9*, 1467 (1970)
2. B. Drerup: Interner Bericht SFB 88/C1 Nr. 11 Münster (1977)
3. Y. Yoshino: Kogaku, Jpn. J. Opt. *1*, 128 (1972)
4. M. Idesawa, T. Yatagai, T. Soma: Appl. Opt. *16*, 2152 (1977)
5. G. Windischbauer, A. Cabaj, G. Keck: In: 1. Jahrestagung der öst. Ges.
    f. Biomed. Technik (19.-22. Mai 1976, Graz)
6. D.M. Meadows, W.O. Johnson, J.B. Allen: Appl. Opt. *9*, 942 (1970)
7. P. Benoit, E. Mathieu, T. Thomas: Opt. Commun. *15*, 392 (1975)
8. W. Jaerisch, G. Makosch: Appl. Opt. *12*, 1552 (1973)
9. G. Keck, G. Windischbauer, G. Ranninger: Optik *37*, 310 (1975)
10. H. Takasaki: Appl. Opt. *12*, 845 (1973)

Problems of Image Evaluation in Orthopedics Using Moiré Figures

G. Windischbauer

Institut für Medizinische Physik, Veterinärmedizinische Universität Wien
A-1030 Wien, Linke Bahngasse 11, Austria

1. Introduction and Survey of Methods

The generation of equal-height contour lines is a promising method for the complete, accurate,and form-fitting description of objects with irregular, threedimensional shapes. This problem exists
in the field of medicine, because traditional measurements of human body forms with tape and calipers are no more sufficient. The
increasing importance of such methods is indicated by the many attempts reported in the past[1]. But in most cases the point coordinates of the body surface had to be extracted before plotting
the contour lines. This procedure was time-consuming or needed a
complex equipment. None of them was accepted in clinical practice.
A great interest arose in all methods, which are capable of generating contour lines as a first "integral" result - without the need
of any analytical handling. One of them, the "Moiré-Topography" of
TAKASAKI,was the starting point of a great number of studies for
the determination of body volume, surface area and shape. Moiré -
contouring is used for the evaluation of the face, of teeth and
feet as well as for whole body photography in radiology and orthopedics[2,3,4,5,6]. Especially in orthopedics there is an increasing demand for description, documentation and evaluation of defi-
nite statements of structural abnormities and obvious asymmetries
of the human body in the treatment of scoliosis[7,8,9,1o].

In these studies three different methods of Moiré-contouring
were used:
the grating-shadow-method, TAKASAKI 197o, [11]
the grating-projection-method,WINDISCHBAUER 1971,MILES 1975, [12,13]
the grating-television-method,WINDISCHBAUER 1976,HORMIERE 1976[14,15]

In all three methods the Moiré-fringes are generated by the
superposition of a periodical master(line)grating and its image on
the object. In the first method this superposition takes place in
observing or photographing the object. In the grating-projection-
method the master grating is superimposed in the image plane of
the objective in a second step. In the grating-TV-method the information about the object(grating) is transformed into a video signal
before the superposition takes place in the electrical path. In
further modifications the video-signal is digitized, fed to a computer and then superimposed numerically on a master grating.Fig.1
shows the essential features of the three contouring methods. The

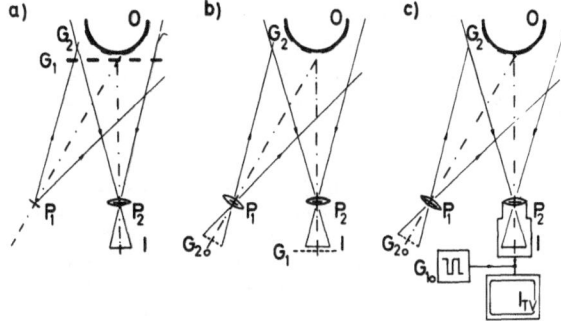

Fig.1 Comparison of three different methods for Moiré-Contouring:
a) grating-shadow-method,
b) grating-projection-meth.
c) grating-TV-method

first one dominates by its simplicity, shadowfree illumination and vanishing of the grating structures. The second method offers a greater viewing area, high quality object gratings and less restrictions of the geometrical arrangement. The grating-TV-method combines the advantages of both and offers the most flexible generation of Moiré-fringes. All three methods generate contour lines of equal height, if some geometrical and optical conditions are met. Thus,the Moiré-contoured image of an object enables an understanding about the body form on the basis of the lines on a map. But are all the problems solved now in measuring body forms? Without any modification these photographs may be used for a raw extraction of shapes in (mass)screening. But for measurements some important parameters must be considered:
the ambiguity of the Moiré-fringes,
the perspective and the image correction,
the performance of the components.
These parameters are limiting the reliable information and the overall accuracy at a reasonable expense.In this paper, the solution for the first problem,the methods for the second and the instruments for controlling will be discussed.

2.   The Ambiguity of Moiré-Figures

Moiré contour lines show no difference on a surface of concave or convexe curvature. The problem is known but all solutions are unsatisfying. An analytical treatment was done by BASSE [16]. Because the algorithm is known, the essential result will be repeated verbally for explaining the new proposal.

The formation of Moiré-fringes may be understood as the correlation of the transmissions functions T_1,T_2 of the two gratings. Contour lines with a non symmetrical profile will be generated by correlating a symmetrical and an asymmetrical transmission function. These lines are capable of discriminating the curvature. But in the grating-shadow-method the correlation of even an asymmetrical grating with its shadow will generate always fringes with a symmetrical profile.

67

2.1 A Modified Grating for the Grating Shadow Method of TAKASAKI

Usually gratings are produced photographically on glass plates or mechanically by stretching a nylon thread over a frame. The transmission functions of such gratings are square angle functions.

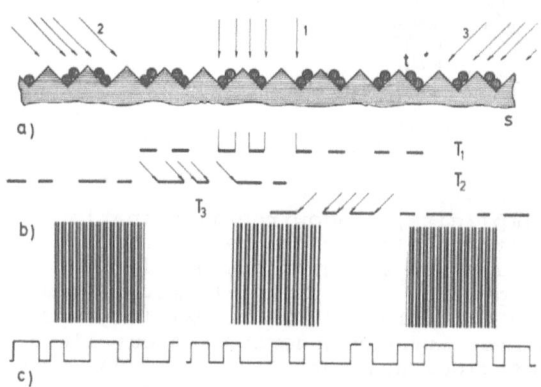

Fig.2 Modified grating for asymmetrical fringe-profile:
a) Mechanical construction principle (t..nylon thread, s.. long srew,T_1..transmission in direction 1)
b) Photographs of a mechanical grating in transparent light
c) Equivalent binary gratings

If the position of the thread is changed in the manner indicated in Fig.2a, a grating is produced, which has a symmetrical transmission function T_1 for rays perpendicular to the grating plane. For inclined rays the functions T_2,T_3 are asymmetrical. Fig.2b shows the photographs of a mechanically constructed grating for three directions. The equivalent binary gratings are drawn in

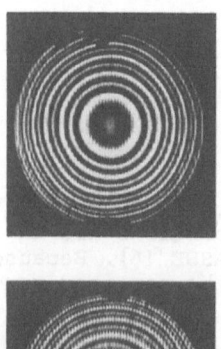

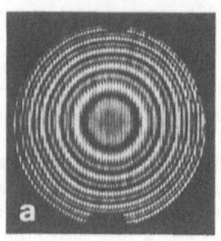

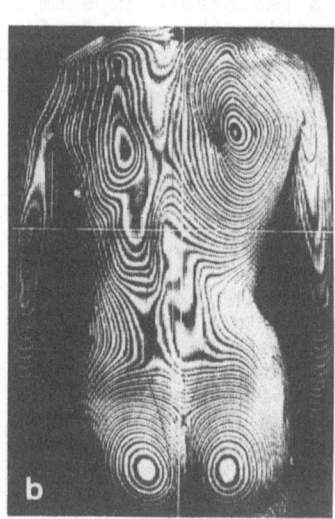

Fig.3 Moiré-contouring with asymmetrical fringe-profiles: a) Two technical objects with a convexe and a concave curvature, b) Moiré-contouring of a girl

Fig.2c. The same effect can be obtained by printing a glass plate
on both sides according to the final transmission of Fig.2a. If
such a grating is used in the basic arrangement of the grating-
shadow-method, the correlation between the asymmetrical (shadow)
grating and the symmetrical (master)grating is performed over the
whole viewing area. Moreover these gratings are suited for shadowfree illumination with two symmetrically arranged light sources.

The formed Moiré-fringes have an asymmetrical profile thus enabling the discrimination of elliptical and hyperbolical points
on the surface. Fig.3a shows two technical objects with a convex
and a concave curvature. Fig.3b shows an orthopedic application.

2.2 Asymmetrical Moiré-Fringe-Profiles for the Grating-Projection-Method and for the Grating-TV-Method

For these two methods the realization of the basic idea is much
simpler.If a grating with an asymmetrical transmission function
(Fig.2c) is projected onto the object, then a symmetrical grating
may be easily superimposed at the image plane. In the grating-TV-
method the projected asymmetrical grating is superposed on a grating electronically generated by a pulse generator. The resulting
Moiré-fringes will show asymmetrical fringe profiles.

3. The Central Perspective and its Correction

In all three methods of Moiré-contouring the centres of projecting
and imaging are in the finite. This implies the changing of the
scale-factor as well as the variation of the distance between two
succeeding Moiré-planes with growing depth of the object. Measurements without correcting these distortions are inadmissible. For
correcting a perspective distortion the centre of projetion must
be known.

   In a first approach all components can be considered to be
ideal especially the point-sources of light and the pupil.
Then single measures may be calculated using simple formulas or
graphical aids for scaling and depht correcting. But every straight
line between two points in the image is defined in a plane including the centre of projection. This means, that cross-sections of
the object, which are drawn parallel in the image,are on inclined
planes(Fig.4).

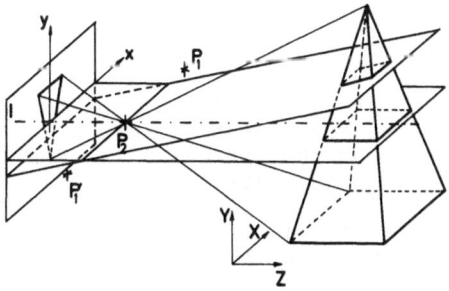

Fig.4   Perspective imaging of an
object.Inclined planes are parallel
cross sections in the image plane.
(P..centre of projection,x,y ...
coord.system of the image plane,
X,Y,Z..coordinates of the object area

Actually the centre of projection is unknown. Its coordinates
are unknown in relation to the image plane (interior orientation)
as well as to the object area (exterior orientation). The image
can only be corrected by the use of control points defined
in the object area and identified in the image.This correction
may be done graphically,optically or numerically, but only in the
case of plane objects [17]. Each Moiré-plane can be corrected if
the control points exist. To obtain these points (at least four
per plane) four straight lines are defined by eight physical points
around the object. If all points of an image have been corrected,
the resulting image can be plotted.It is the orthogonal projection
of the object. It is only this plot which permits direct measurements in the image - at the expense of an analytical point-to-
point-prehandling!

4.   The Performance Characteristics of the Components

Many details are dicussed in other studies. But two essential
points have been overlooked: the camera and the adjustment. Both
must be seen critically. In most cases non-metric cameras are
used. Because of the instability of the interior orientation its
adjustment has to be checked .Fig.5 shows the apparatus we use
for Moiré-Contouring in orthopedics (object area:9ox6ox3o cm,re-
solution in depth: 2-2,5mm, asymmetrical Moiré-fringe-profile).

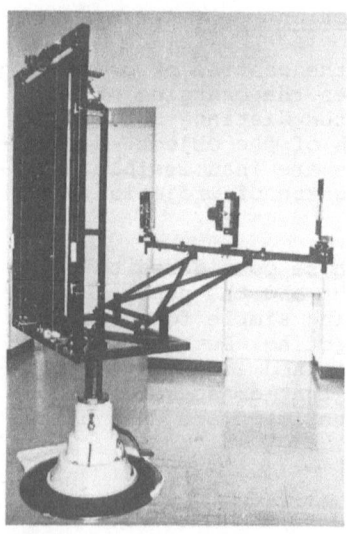

Fig.5   Apparatus for Moiré-
Contouring in Orthopedics

The final overall accuracy is unknown but there is good hope to
deviate less than 1mm. For control of the adjustment, we use a
metal bench which rotates precisely around the optical axis. For
adjustment the grating and the control points are scanned mechanically. For control the test object will be easily replaced with
high precision. The goal is to simplify the algorithm for image
correction by a proper adjustment. Measurements should be done by
using only a ruler and a pocket calculator.

5. Summary

In the studies of human body form three different methods of
Moiré contouring are used. All of them are capable of generating
Moiré-fringes with an asymmetrical fringe profile. Thus, the discrimination of convex and concave curvature of a surface is
possible. All Moiré-methods are suited for screening and raw body
form extraction. But for measuring purpose the perspective correction of the image and the control of the adjustment is necessary.

1. R.E. Herron: Stereophotogrammetry in Biology and Medicine.
   Proc.XII Congr. Int. Soc. Photogr. (1972) Ottawa.
2. E. Kanazawa, H. Kamiishi: Evaluation of Facial Osteotomy with
   the aid of Moiré Contourography. J. Max.-fac. Surg. 6 (1978)
   233.
3. W. Holler: Vergleichende Untersuchungen der oberen ersten
   Molaren bei Wolf (canis lupus lupus), Haushund (canis famili-
   aris-Dobermann) und Goldschakal(canis aureu) mit Hilfe der
   Moiré-Topographie. Säugetierkundliche Mitt. 26 (1978) 131.
4. G. Windischbauer, G. Keck, A. Cabaj: Moiré-topographische
   Verfahren in der Medizin: Ein Podometer zur Dokumentierung
   von krankhaften Fußstellungen. Orthop.-Technik (1977) 7.
5. W. Binder, A. Cabaj, K.H. Kärcher, G. Windischbauer: Her-
   stellung von Gewebeausgleichsfiltern für große Flächen bei
   Hodgkin-Bestrahlung mit Hilfe der Moiré-Technik. Strahlenth.
   153 (1977) 82.
6. H. Neugebauer, G. Windischbauer: Die Moiré-topographische
   Dokumentation von Haltungsformen und Wirbelsäulenerkrankungen.
   Biom. Technik. 23 (1978).
7. B. Drerup: Eine Apparatur zur Dokumentation von Erkrankungen
   des Haltungs- und Bewegungsapparates durch Moiré-Topographie.
   Int. Bericht Univ. Münster (1977).
8. B. Drerup: Anwendung der Moiré-Topographie zur Diagnose und
   Dokumentation von Fehlbildungen des Rumpfes. Z. Orthp. 116
   (1978) 789.
9. H. Neugebauer, G. Keck, G. Windischbauer: Die Moiré-topographische Dokumentation der Wirbelsäule und des Rückens. Öst.
   Ärztez. (1978) Nr. 12.
10. St. Willner: Moiré-Topography: Method for three dimensional
    study of asymmetries of the human back. Inf. Fa. Camp.,
    Helsingborg, Sweden.
11. H. Takasaki: Moiré-Topography. Appl. Opt. 9 (1970) 845.
12. G. Windischbauer: Oberflächenvermessung des menschlichen und
    tierischen Körpers mit Hilfe von Moiré-Figuren. Proc. 1st
    Biophys. Cong. 6 (1971) 512 Baden/Vienna.
13. C.A. Miles, B.S. Speight: Recording the shape of animals by
    a Moiré method. J. Phys. E. 8 (1875) 773.
14. G. Windischbauer, A. Cabaj, G. Keck: TV-Moiré-Topography.
    Wiss. Berichte Tag. Biomed. Techn. in Österr. (1976) 177,Graz.
15. J. Hormiére, E. Mathieu: Reference plane orientation using
    optoelectronic methods in Moiré contour lining. Opt. Comm.19
    (1976) 37, Nr.1.
16. G. Basse: Moiré-Methoden zur Erzeugung von Höhenschichtlinien
    mit asymmetrischen Profilen. Diss. Univ. Hamburg (1967).
17. R. Finsterwalder, W. Hofmann: Photogrammetry. Walter de
    Gruyter & Co, Berlin (1968)

IV. Holography in Biology

Autocorrelation of Diatoms as a Function of Depth of Focus

J.K. Partin, S.P. Almeida, and H. Fujii

Department of Physics, Virginia Polytechnic Institute and State University
Blacksburg, VA 24061, USA

1. Introduction

The use of matched spatial filters in the identification of diatoms on 35mm
transparencies has been previously reported [1,2]. In these papers the diatoms were said to play an important role in water pollution monitoring. Also
that their identification by humans requires a trained taxonomist and is a
tedious process. While the use of spatial filtering has been shown to work,
one would like to avoid, if possible, the use of transparencies as input to
the hybrid optical processor and, instead, work directly with the microscope
slides. This method, however, introduces new problems. One problem being
speckle. That is, the coherent illumination of a microscope slide introduces
a speckle-noise background that degrades the diatom image. Preliminary results on the speckle problem have already been obtained and reported [3].
Another problem is that of depth of focus. A study of this problem and how
the autocorrelation signal changes as a function of depth of focus is now
presented.

2. Microscope Optical Processor

Figure 2.1 shows the schematic of the microscope optical processor. The microscope slide containing the diatoms is illuminated by a Helium-Neon laser.
The diatom of interest is selected out by means of an iris. This signal is
then Fourier transformed onto the photographic plate where it is interfered
with the plane reference wave. A matched spatial filter is thus made on the
film plate for later use in the identification of the same diatom. Vidicon
I is used to observe the autocorrelation signal as a function of depth of
focus. In addition, 256x384 points are digitized into 64 gray levels and
plotted. Vidicon II serves to monitor the microscope input diatom under
investigation. A photograph of the optical system is shown in Fig 2.2.

3. Autocorrelation Signals and Depth of Focus

The study of autocorrelation signals as a function of depth of focus is important because of the high probability of finding misfocused specimens on
a slide. Such a study was undertaken on fifteen different diatoms. The
autocorrelation signal was first measured for the diatom in focus. Then
the specimen was misfocused by 4μ, 8μ, and 22μ in both the forward and back-
ward positions. This was done for a magnification of 25X and 40X. In each

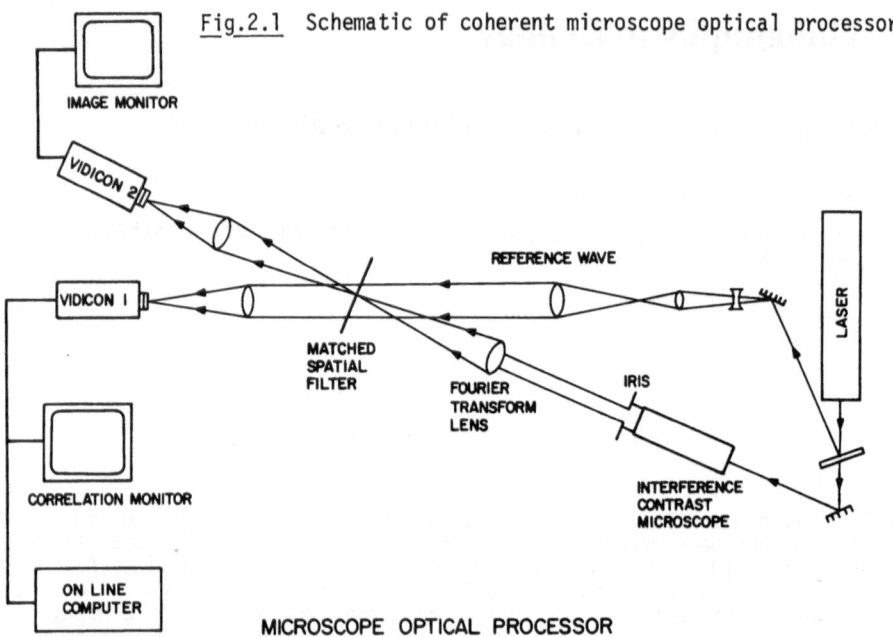

Fig.2.1 Schematic of coherent microscope optical processor.

IMAGE MONITOR

VIDICON 2

VIDICON I

REFERENCE WAVE

LASER

MATCHED
SPATIAL
FILTER

FOURIER
TRANSFORM
LENS

IRIS

CORRELATION MONITOR

INTERFERENCE
CONTRAST
MICROSCOPE

ON LINE
COMPUTER

MICROSCOPE OPTICAL PROCESSOR

Fig.2.2 Photograph of the optical system

case where the specimen was in focus the signal was normalized to the same
value. Then the autocorrelation signals obtained for all 15 diatoms at each
misfocused step were averaged and plotted. Results obtained for two of the
fifteen diatoms are presented in Fig.3.1a-h and in Fig.3.2a-h. These figures
show a detectable signal to about eight microns misfocus. The averaged autocorrelation signals for all fifteen diatoms are shown in Fig.3.3 for both
25X and 40X magnification.

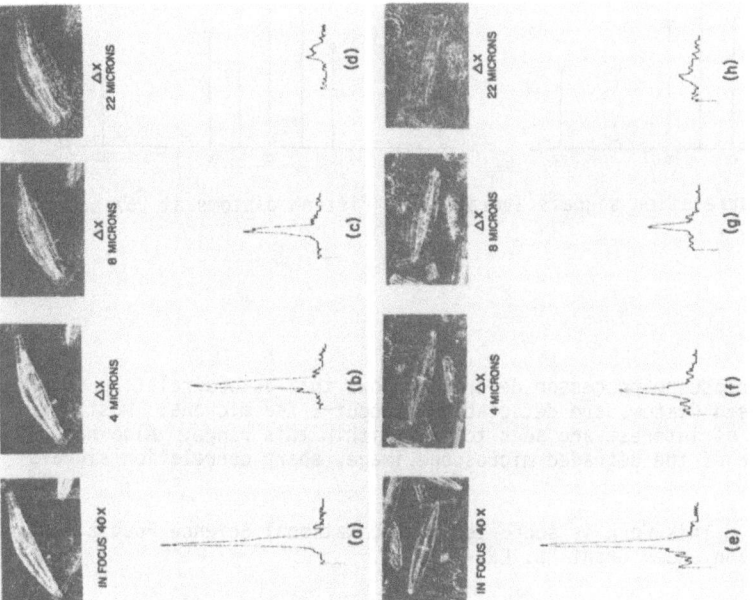

Fig.3.2a-h. Pictures taken through the microscope of two diatoms misfocused in the forward direction. Auto-correlation signals as a function of depth ΔX of focus for two diatoms at 40X

Fig.3.1a-h. Pictures taken through the microscope of two diatoms misfocused in the forward direction. Noise background in the pictures is due to coherent illumination and speckle. Autocorrelation signals as a function of depth ΔX of focus for two diatoms at 25X

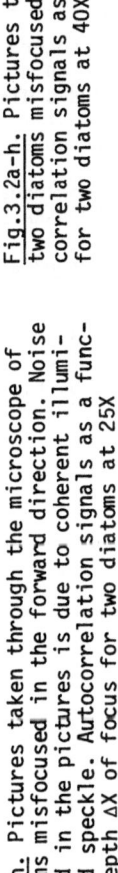

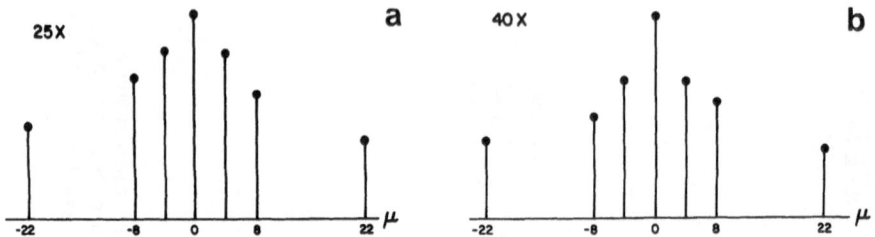

Fig.3.3. Autocorrelation signals averaged for fifteen diatoms at 25X and 40X magnification.

4. Conclusion

The coherent microscope processor described shows that autocorrelation signals on misfocused diatoms are detectable to about $\pm\Delta X=8$ microns. Most misfocused diatoms of interest are seen to fall within this range. Also noted is that in spite of the degraded microscope image, sharp correlation signals were obtainable.

Acknowledgement   This work is supported by the National Science Foundation, PFRA/ASRA Division under Grant No. ENV-7710184.

References

1. S. P. Almeida, J. K. T. Eu: Appl. Opt. 15, 510-515 (1976).
2. S. P. Almeida, J. K. T. Eu, P. F. Lai: IEEE Trans. on Meas. and Instr., IM-26, No. 4, 312-316 (1977).
3. S. K. Case, S. P. Almeida, W. J. Dallas, J. M. Fournier, App. Phys. 17, 287-293 (1978).

A Simple Setup for Using Holographic Interferometry in Studies on Seeds

J.J. Lunazzi and L. Wickert[1]

Instituto de Fisica, Universidade Estadual de Campinas
13.100-Campinas - SP - Brazil

1. Introduction

In order to enable the use of holographic interferometry outside a specialized physics laboratory simple and stable equipment is often required. Many efforts have previously been made
on this subject [1] [2] [3] [4], including those of WILLIAMS
[5], and FOX and PUFFER [6], who succeeded in applying holographic
interferometry to the field of plant physiology.
    Low cost, ease of handling and an excellent degree of stabili-
ty are characteristics of the sandbox holographic system shown in
Fig. 1.

Fig. 1 Sandbox set-up for holographic interferometry, including
the laser beams, the paths of which, starting from the laser,
can be seen up to the positions of the reference mirror and of the
object. L = laser, M = mirrors (see text), SP = spatial filter,
O = object, H = holographic plate.

2. Description of the system

The system [7] presented in Fig. 1 is based on a previous arrangement [8] that was adapted to work in a sandbox of dimensions

[1]On leave from the University of Passo Fundo, on scholarship from
CAPES - Ministry of Education.

$30 \times 40 \times 90$ cm^3. The beam from a 0.4 mW He-Ne laser impinges on a first surface mirror M_1 that directs it to a second surface mirror M_2 from which we obtain three parallel reflected beams. The central beam is the more intense one and it is focused by a microscope objective together with one of the lateral beams. At the focal point both beams are spatially filtered by a single pinhole. Thus, we obtain two clean beams: an intense one for illumination of the object and a weaker reference beam. Up to this point both beams have traversed the same optical elements, so that any phase shift affecting one of them will affect the other as well. The difference in phase between both beams at the plane of the holographic film is therefore only little affected by disturbances within this part of the system.

The object and the film holder are supported on the same iron base by means of magnetic pieces, and it is possible to add a third surface mirror M_3 for better orientation of the reference beam.

All the elements were finally mounted on thick plastic tubes allowing easy positioning. In this way, the alignment of the set-up is highly simplified. Care must only be taken of the fine adjustments in two angular degrees of freedom of each element. Even for this fine adjustments, the little backlash that is present can be considered to be less important than that of most adjusting knobs on standard holographic systems. As a consequence, the delicate task of aligning the spatial filter for two beams simultaneously can be easily performed.

Special attention is given to the description of the spatial filter: it was constructed with three square metal plates each 1 cm thick, and assembled by sliding a microscope objective into a proper hole made through the first plate. A steel spring was then fixed around that hole. A pinhole, 25 µm in diameter, was glued to the first turn of the spring after centering its position. Two fine standard screws were finally used in the last plate for adjusting the pinhole position in the optical axis. They were aligned in mutually perpendicular directions acting on the last free turn of the spring, resulting in the required micrometric positioning accuracy. Finally, the spring was wrapped in foam rubber within a plastic tube in order to dampen its natural modes of vibration.

The intensities at the plane of the film and the transmittance of the hologram were determined by using a simple photometer made of one cadmium-sulfide photocell whose resistance was measured with a common electrical tester. A simple calibration in arbitrary units of luminous intensity is just required for this photometer. Standard quality mirrors were employed and the required mechanical isolation was achieved by supporting the system on four pneumatic tubes.

3. Testing of the system

By means of the described procedures we obtained a ratio 12:1 between the beam intensities at the exit of the spatial filter and, for the array of white beans showed as holographic objects in Fig. 1, the ratio was 3:1 at the film plane. Exposure time was 15 s for each holographic record. Sheets of Kodak SO-173 holo-

Fig. 2 Holographic double-exposure interferograms showing the surface deformations of beans caused by breathing over them.

graphic film were fixed in a rigid plastic support made of two transparent plates of acrylic material. Initially, double-exposure holograms were taken of the array of beans by just breathing over them once between the exposures. As a result of these preliminary investigations clearly visible interference fringes were obtained, demonstrating rigid body movements of each bean as well as local surface deformations.

Many double-exposure holograms of rigid objects were made by using Kodak holographic plates of the 131-02 type, for testing the stability of the system. Interference fringes were observed only at intervals of two hours between the exposures. The directions of the fringes was mainly horizontal demonstrating a flexure of the wooden base of the box, a problem that can be avoided by using a metal base. The system proved to be very stable, even in cases when big machines were used in the building and many different vibrations were present during the exposures. No parasitic fringes could be detected at exposure separation times of five minutes.

4. Description of the investigations

Surface deformations in the region denominated as Hilum of seeds of beans (Phaseolus vulgaris L.) were investigated holographically using the described system. Humidity changes were obtained by means of water evaporating from an opened little box forming a water surface of 9 cm^2, which was placed at a distance of 4 cm under the seeds a few minutes before the exposures.

The interpretation of the interference fringe pattern was simplified by the procedure used to fix the seeds: the lower part of each seed, just under the Hilum, was glued with a transparent epoxy material to a glass plate. Thus, the body of the seed is fixed and only its free surface is affected by humidity changes. The glass plate with the seeds was then vertically positioned, so that the normal to its surface bisected the angle between the illumination and observation direction, that is, in the direction of the sensitivity vector K [9]. This direction can be regarded, with little error, as constant over the region of observation for each seed. A white painted metal surface was placed after the seeds as a reference object for checking the stability of the system. The fringe pattern can then be interpreted by considering that each fringe corresponds to points of equal displacement in the direction of the sensitivity vector.The displacement characterized by two adjacent fringes corresponds to, typically, 0.4 µm. Displacements perpendicular to the direction of the sensitivity vector (in-plane displacements) are not detected in this way. Zero order fringes were determined by searching for fixed fringes when looking at the virtual image from every per-mitted direction. The first zero order fringe could always be determined at the glued part of a seed. The fringe pattern was finally observed and photographed by projecting the real image through a small area of the hologram. Very simple symmetric fringe patterns (see Fig. 3) could be detected, corresponding to a 5 min. interval between both exposures. These patterns consisted of the first zero order fringe and a single closed dark fringe followed by another closed bright fringe. The latter, having an annular form, corresponds to a new zero order fringe that indi-cates the immobility of that annular region, mainly in its lower and upper part.

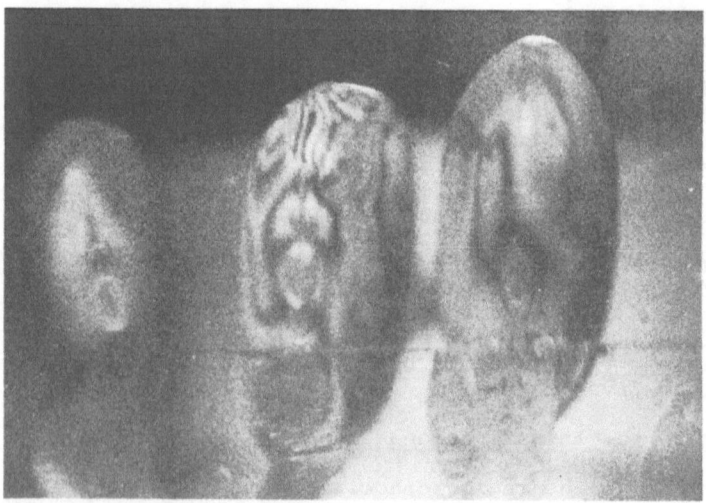

Fig. 3 Fringe pattern on the Hilum of Phaseolus vulgaris L

Yet some displacements are detected at the central part of the surface of the Hilum since the last bright fringe darkens there or even a new bright fringe appears in the form of a central spot.

As a final example an investigation on seeds of maize carried out with this set-up is mentioned. A single dark fringe was ob-served surrounding the part of the tegument covering the embryo when the seeds of maize (Zea Mais L.) were submitted to the same procedures as described above.

5. Conclusions

A simple and stable holographic set-up can be built with a minimum of elements. The most expensive parts of the system are a standard low-power He-Ne laser, and a spatial filter consisting of a microscope objective and a pinhole. The described system is very reliable and can be easily constructed. It turned out to be a powerful tool for holographic interferometric investigations on seeds exposed to humidity changes. The micrometric deformations arising can be analyzed by simple procedures. Thus, it is possible also for non-skilled people to apply with ease holographic techniques in biology.

References

1 Collier,R.J., Burckhard,C.B., Lin,L.H., "Optical Holography" Academic Press 1971.

2 Taylor,C.E., Amer.J.Phys. V.39, p.417 (1971).

3 Wise,J., Amer.J.Phys. V.40, p.1866 (1972).

4 Porter,A.G., George,S., Amer.J.Phys. V.43, p.954 (1975).

5 Williams,G.T., Nature V.215, p.1170 (1967).

6 Fox,M.D., Puffer,L.G., Plant Physiol. V.60, p.30 (1977).

7 Wickert,L., Thesis, Universidade Estadual de Campinas, 1978.

8 Stetson,K.A., Powell,R.L., J.O.S.A. V.56, p.1161 (1976).

9 Stetson,K.A., J.O.S.A. V.64, p.1 (1974).

V. Holography in Radiology

Holography in Radiology

H. Weiss

Philips GmbH Forschungslaboratorium Hamburg
D-2000 Hamburg 54, Rep. Fed. of Germany

1. Introduction

At the first glance there seems to be no relation between holography and radiology. Holography uses visible light to store and
retrieve three-dimensional information by interference of coherent waves, a possibility X-ray imaging symstems lack because of the incoherent nature of X-rays. X-rays are not reflected or diffracted by lenses, mirrors or gratings like electro-
magnetic waves of larger wavelength. But this draw-back of X-
rays actually gives rise to usage of holography and optics in
a synthetic way of postprocessing X-ray pictures to obtain the
flexibility of optics like it has been done very successfully
in the field of radar technique [1].

The only physical change an X-ray beam suffers from penetrating
the body is loss in intensity because of absorption. Hence a conventional X-ray image is shadow casting the absorption properties of a three-dimensional object onto a two-dimensional plane.
No information on the depth of the details within the object is
obtained.

The first attempt in radiology to image defined layers of the
object was made by the invention of tomography, a very often
used technique.

In tomography (Fig.1) exactly one layer TL of the 3-d object O
is imaged in focus by a synchroneous movement of X-ray tube T
and film F along the so-called blurring path. All other layers
are blurred with a point-spread function according to the blurring curve. All information outside the tomographic plane is
"wasted" and cannot be retrieved. All attempt in the past years
concentrated on improving - from the point of view of information theory - this ineconomic imaging method.

2. Three-Dimensional Back Projection

Already very early after introduction of tomography the idea of
discretizing the blurring path came up [2]. There a set of radiographs is taken from different views and afterwards by back

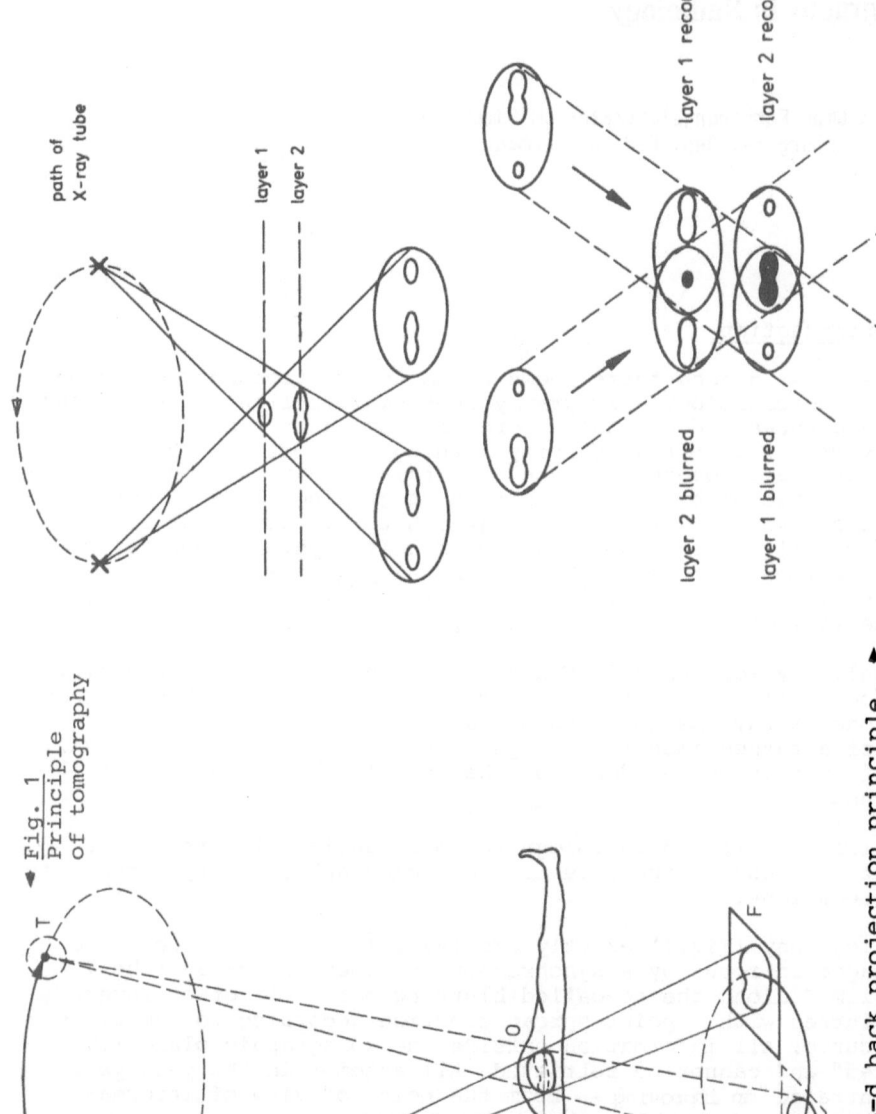

path of
X-ray tube

layer 1
layer 2

layer 1 reconstructed

layer 2 reconstructed

layer 2 blurred

layer 1 blurred

Fig. 1
Principle
of tomography

Fig. 2a,b. 3-d back projection principle. ▲
(a) Decoding step. (b) Reconstruction step

projecting these two-dimensional projections the three-dimensional object is reconstructed (Fig.2).

The principle of back projection can be verified in two ways. Either one is generating certain layers (Fig.2) by inserting in the corresponding planes a screen for displaying tomograms or a virtual image is reconstructed resulting in a quasi-3-d picture of the object. Both methods were implemented with holographic post-processing of the primary X-ray pictures [3], [4]. In Fig. 3 the holographic tomosynthesis principle is sketched. From the primary X-ray pictures obtained along a circular movement of the X-ray tube flashed at discrete positions a ring of holograms is made (Fig.3a).

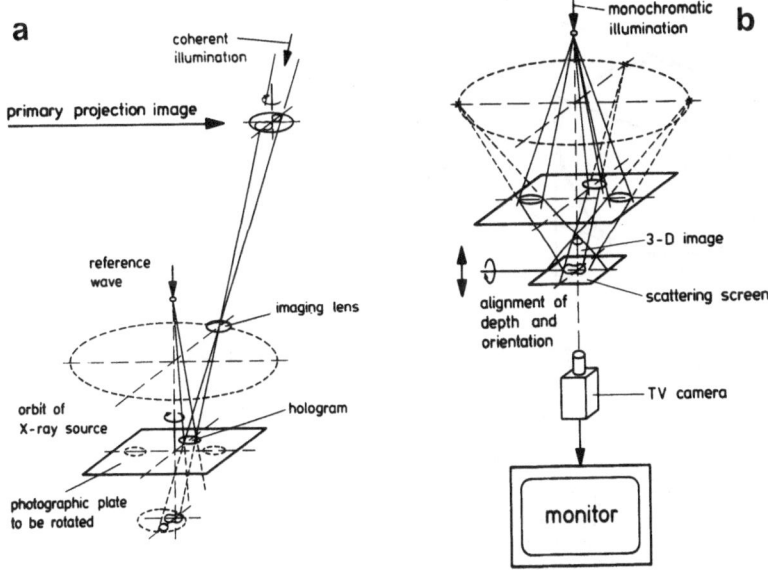

Fig.3 Holographic back projection technique
        a) producing the holograms
        b) reconstructing a real 3-d image of the object

By illuminating this composed hologram from each of the holograms one view on the object is back projected and all these projections are interfering in the reconstruction volume where a real 3-d image of the object is generated and arbitrary tomograms can be displayed e.g. using a TV-chain. Holography offers here its capability of storing pictures together with their directions from which they were received. Problems in this back projection are the transfer of grey tone pictures of high dynamics into spatial frequencies within the hologram, which cannot be performed completely [5], and the introduction of sophisticated holographic techniques into medical routine environment.

3. Filtered Back Projection: Computerized Tomography

The invention of computerized tomography [6] brought a new dimension of accuracy into radiology. At the first time it was possible to image not only high contrast objects like bones or contrast media given into the body, but also to distinguish different kinds of tissue. This is due to two new techniques used, namely very accurate complete measurement of the object absorption and very accurate reconstruction of the absorption values by computer technique.

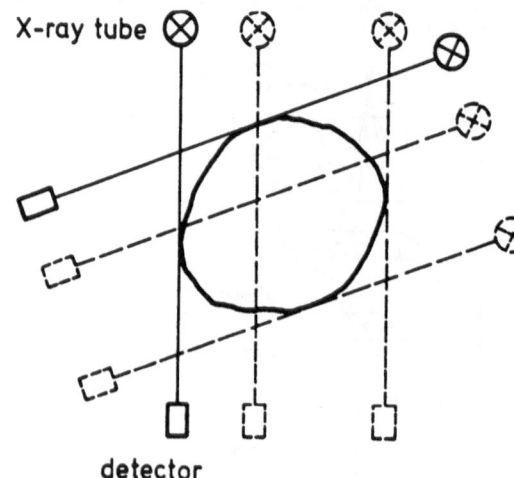

X-ray tube

detector

Fig. 4. Principle of computerized tomography (recording step)

Fig.4 shows the scanning of a two-dimensional body slice by moving a pencil X-ray beam and its detector across the section linearly followed by a rotation by a small angle, again scanning etc. until the full cycle of 180° is covered. The integral absorption profiles are measured and then from this one-dimensional projection the absorption distribution within the slice is calculated.

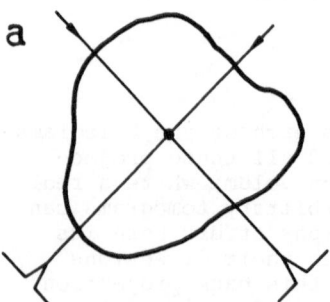
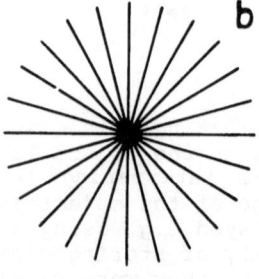

a

b

Fig.5 Principle of reconstruction in computerized tomography
a) projecting one point of the object
b) back projecting these point-projections

The reconstruction process principally can be explained from Fig.5. In Fig.5a only the projection of one point within the object is shown resulting in a set of about 180 δ-functions - according to 180 projections -. By only back projecting these δ-functions one gets a distribution shown in Fig.5b, which approximates a function $1/r$, if an infinite number of continous projections is assumed. Because of the linear shift-invariant behaviour of the total imaging system this function $1/\gamma$ is the point-spread function of the reconstruction procedure. Obviously the back projected image ρ_B of an absorption distribution ρ within the body slice is then given by the equation

$$\rho_B = \rho * \frac{1}{r},\tag{1}$$

* denoting convolution. The task now is to deconvolve (1) in order to retrieve the absorption values ρ. The back projected absorption ρ easily can be obtained from the projections. Principally the convolution can be performed by Fourier-transforming (1)

$$F(\rho_B) = F(\rho) \cdot F(\frac{1}{r}) = F(\rho) \cdot \frac{1}{\rho}\tag{2}$$

Since the Fourier transformation of $F(\rho_B)$ is known only up to a cut off frequency, an apodizing function $A(\rho)$ has to be introduced such that than from (2)

$$F(\rho) = F(\rho_B) \cdot \rho \cdot A(\rho)\tag{3}$$

results. By inverse Fourier transformation we get the final absorption values

$$\rho = F^{-1} [F(\rho_B) \cdot \rho \cdot A(\rho)]\tag{4}$$

The inverse Fourier transformation can be performed more or less partially, getting from (4)

$$\rho = \rho_B * F^{-1} [\rho \cdot A(\rho)] .\tag{5}$$

Both deconvolution techniques can be performed also using optics [7], [8]. Eq. (4) requires optical Fourier transforming the back projected values ρ_B, multiplication in the Fourier plane by the filter function $\rho \cdot A(\rho)$ and again Fourier transformation. If this is done with a coherent optical set-up, one is confronted with all troubles usually encountered in coherent processing. Using Eq. (5) one operates in the image space. The convolution kernel $F^{-1}[\rho \cdot A(\rho)]$ is a real function; the convolution with the positive valued function ρ_B can be performed in two steps, first by convolving with the positive and then with the negative part of the kernel, and afterwards subtracting both terms. But at this point it has to be mentioned that any optical implementation of computerized tomography reconstructions is confronted with three severe problems. The first is the conversion of electrically measured projection values into optical transparencies, because usage of X-ray film in the recording step instead of detectors is very problematic because of its limited signal to noise ratio and reproducibility of the exposure curve. The second problem is the requirement of a very accurate deconvolution and the third is the inflexibility of optics to all preprocessing of the data like calibration and correction for the polychromatic X-ray beam.

4. Coded Aperture Imaging

The very special synthetic aperture of using discrete positions along a circle for imaging three-dimensional objects as explained in chapter 2 can be generalized to the coded aperture method.

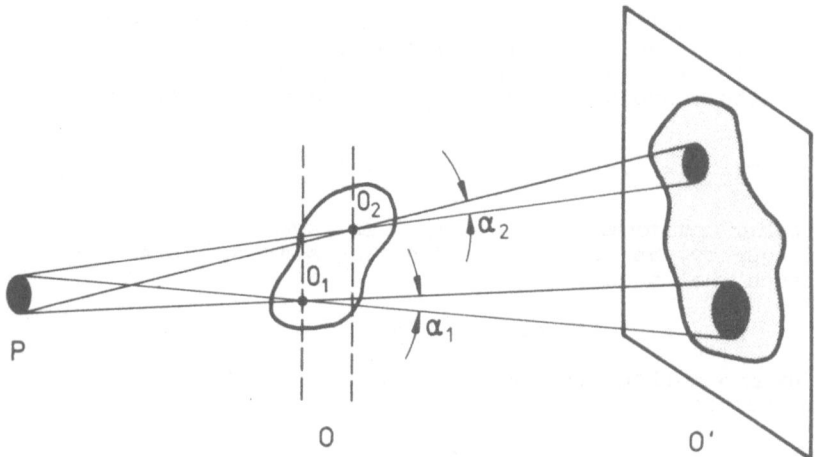

Fig.6 Principle of coded aperture imaging

An extended source P irradiates a 3-d object O consisting of layer O_1, O_2 etc. casting a shadow O' onto a 2-d plane. Due to the different depths of points in layers O_1, O_2 the shadow of P differs in scale. The image O' is given by

$$O' = \sum_i O_i * P_i, \tag{6}$$

the summation taken over all planes, and P_i the corresponding point-spread functions different only in scale. Generally the problem of retrieving the plane O_1 from the coded image O' is solved by convolving with a propriate function Q_1

$$O'' = \sum_i O_i * P_i * Q_1$$

$$= O_1 * P_1 * Q_1 + \sum_{i \neq 1} O_i * P_i * Q_1. \tag{7}$$

In order to be $O'' = O_1$ the equations

$$P_1 * Q_1 = \delta \qquad (8)$$

$$P_i * Q_1 = 0, \; i \neq 1 \qquad (9)$$

should hold, δ denoting a Dirac-function. Requirement (8) defines the quality of the layer O to be reconstructed and requirement (9) characterizes the influence of neighbouring layers O_i, $i \neq 1$, on the reconstructed tomogram. Little effort up to now has been made in order to approximate (9) by choosing appropriate functions Q_1. Actually artifacts caused by neighbouring layers could only be avoided by applying techniques like computerized tomography, but here in three dimensions! The problem of finding P_1 and Q_1 in such a way that (8) is fulfilled is difficult enough taking years of efforts of numerous authors. Concentrating only on one plane O of the object we have to solve equation

$$O' = O * P \qquad (10)$$

for the function δ.

a) Fresnel-Zone Plate

Choosing as pupil function P a Fresnel-zone plate consisting of concentric rings of irradiating and non radiating material each point of the object is imaged into a Fresnel-zone shadow. After developing the film and illuminating this transparency with a plane coherent light wave one obtains an infinite series of real and virtual foci (Fig.7) together with a dc-term of 0. order.

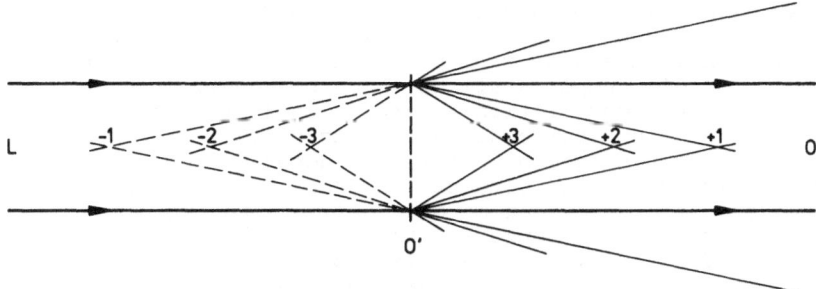

Fig.7 Illuminating a Fresnel-zone transparency with a plane wave

In this case the function Q in

$$O'' = O * P * Q \qquad (11)$$

is the free space itself. Of course attempts have to be made for blocking all foci except one and the dc-term. One of the best approaches to this problem is the off-axis zone plate [9], which has its center outside the image axis and here different foci are imaged in different directions where the blocking conveniently can be made. Since each point of the object is imaged into a Fresnel-zone shadow the developed film plate is a kind of incoherent hologram where all the ring structures are incoherently interfering. This kind of coded aperture was proposed first for γ-imaging with radioactive isotopes because there the Fresnel-zone plate can be realized as a "passive" aperture consisting of rings of lead and γ-ray translucent material. Constructing an "active" X-ray source in form of a zone plate is quite difficult. Hence in radiography other kinds of apertures are used:

b) Point-Distributions

Apertures of a point structure more easily can be realized either by a multi-pinhole camera [10] in γ-ray imaging or as an array of X-ray tubes [11] in X-ray imaging.

Starting from (11) the task is to find two point-distributions P and Q such that

$$P * Q \approx \delta \tag{12}$$

is best fulfilled. The simplest way is to choose for the decoding code Q the negative coding distribution itself, giving

$$P \circledast P \approx \delta, \tag{13}$$

i.e. the autocorrelation of the point-distribution is to approximate a δ-function. For distribution consisting only of +1's, the best in the sense of (13) are the so called nonredundant point-distributions NRPD [12].

In Fig.8 two examples of NRPD's are given. The mathematical property of a NRPD is that the autocorrelation has a peak of intensity n, if n is the number of points in the distribution, surrounded by $n(n-1)$ subpeaks all of which have only intensity 1. Since the autocorrelation according to (11) is the point-spread function of the total coding-decoding system, the side peaks contribute to the "noise" within the decoded image of intensity n. There were many attempts in order to increase the signal to noise ratio. They can be categorized into two directions. The first one is improving the decoding while using an NRPD in the coding step, the other is looking for coherent codes [15] containing also negative valued points.

Improvement in decoding e.g. can be achieved in the Fourier space. There [11] writes

$$F(O'') = F(O) \cdot F(P) \cdot F(Q), \tag{14}$$

and obviously the function $F(Q) = 1/F(P)$ gives a true reconstruction of the object O after inverse Fourier transformation. The general concept of optical inverse Fourier filtering using

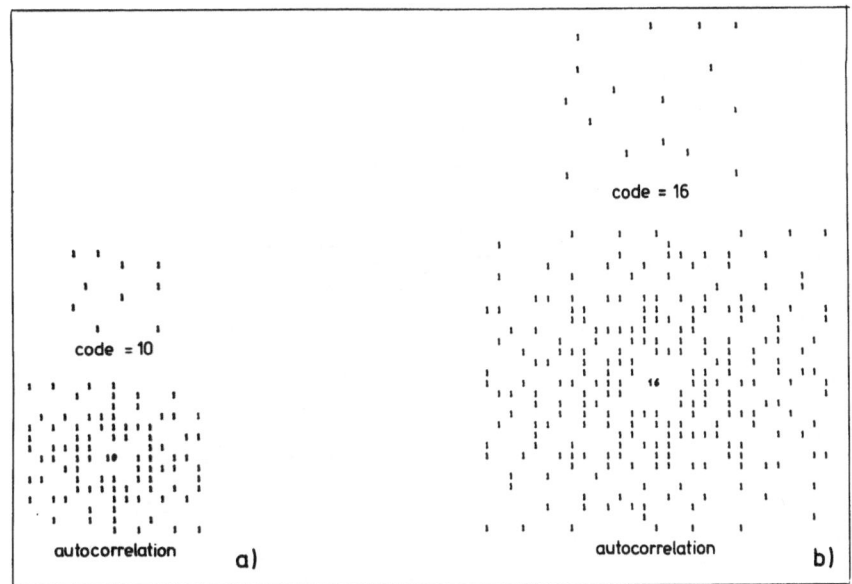

Fig.8 NRPD's and their autocorrelations

coherent light and a hologram for verifying the complex filter function is known since years [13].
The difficulties in producing the inverse Filter 1/F(P), due to the high dynamic range of this function, recently could be overcome by a logarithmic recording of the filter [14], and this technique now gives hope for coming a step further.

The method of using coherent codes also can be implemented by optical means separating the deconvolution in a negative and positive part like in the computerized tomography implementation. This holds for using a coherent code P already in the coding step [15] as well as for methods using an incoherent code for coding and a coherent one for decoding [16], [17]. In both cases optimal codes have been found.

The problem arises whenever these algorithms should be optically implemented due to the necessity of subtracting images, where noise is increased and differentiation effects can occur.

But nevertheless all radiographic imaging systems are linear shift-invariant systems where a deconvolution always can be tried by optics either in the Fourier space or in the image space, either applying holographic techniques for generating Fourier filters or other optical components for processing the primary radiographs. However, opticians never should forget that images with low spatial resolution nowadays easily can be handled by computer techniques.

References

1. Cutrona, L., J. Leith, E.M. Leith, L.J. Porcello, and W.E. Vivian, Proc. IEEE 54, 1036 (1966)
2. Ziedses des Plantes, B.G., Acta radiol. 13, 182 (1932)
3. Groh, G. and M. Kock, Röntgenblätter 9, 451 (1971)
4. Kock, M. and U. Tiemens, Opt. Comm. 7, 260 (1973)
5. Clausen, C. and U. Killat, Med. Phys. 5, 181 (1978)
6. Hounsfield, G.M., Brit.J.Radiol. 46, 1016 (1973)
7. Peters, T.M., IEEE Trans. Biomed. Eng. BME 21, 214 (1979)
8. Edholm, P., Acta Radiol. Diagn. 18, 126 (1977)
9. Barrett, H.H., P.T. Wilson, G.D. De Meester, and H. Scharfman, Optic. Engng. 12, 8 (1970)
10. Groh, G., G.S. Hayat, and G.W. Stroke, Appl. Optics 11, 931 (1972)
11. Klotz, E. and H. Weiss, Opt. Comm. 11, 368 (1974)
12. Golay, M.J.E., J. Opt. Soc. Am. 61, 272 (1971)
13. Stroke, G., R.G. Zech, Phys. Lett. 25A, 89 (1967)
14. Dallas, W.J., R. Linde, and H. Weiss, Optics Lett. 3, 247 (1978)
15. Weiss, H., E. Klotz, R. Linde, G. Rabe, and U. Tiemens, Optica Acta 24, 305 (1977)
16. Fenimore, E.E. and T.M. Cannon, Appl. Opt. 17, 337 (1978)
17. Dallas, W.J., to be published in Opt. Comm.

Flashing Tomosynthesis

H. Weiss, E. Klotz, and R. Linde

Philips GmbH Forschungslaboratorium Hamburg
D-2000 Hamburg 54, Rep. Fed. of Germany

Abstract

Flashing tomosynthesis is a new tomographic method in radiology consisting of two steps. In the recording step the 3-dimensional object is imaged by flashing an array of X-ray tubes simultaneously on one image plan resulting in a coded image of the object. In the second post-processing step this coded image is decoded by optical means displaying arbitrary layers.

1. Principle of Flashing Tomosynthesis

The coded aperture imaging system of flashing tomosynthesis (Fig. 1) aims at a new tool in X-ray diagnosis. In the first step the three-dimensional object is imaged by an array of X-ray tubes resulting in a coded image. This coded image is post-processed in a second step to reconstruct arbitrary layers of the object. Although analog electronic or digital implementations of the decoding algorithm is possible [1], optical deconvolution in this case seems to be the most convenient hardware solution. Whereas the first implementation was performed using holography [2] meanwhile a decoding set-up using only lenses was constructed, a technique quite adapted to hospital environment.

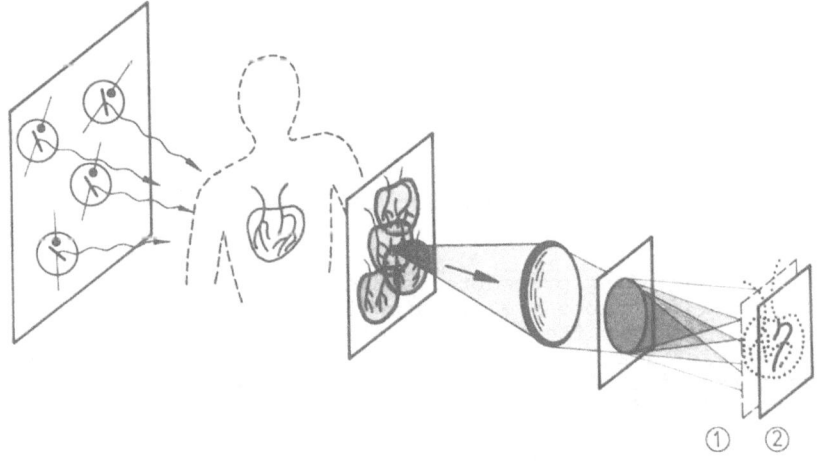

① ②

Fig. 1   Principle of flashing tomosynthesis

Fig. 2 Recording step

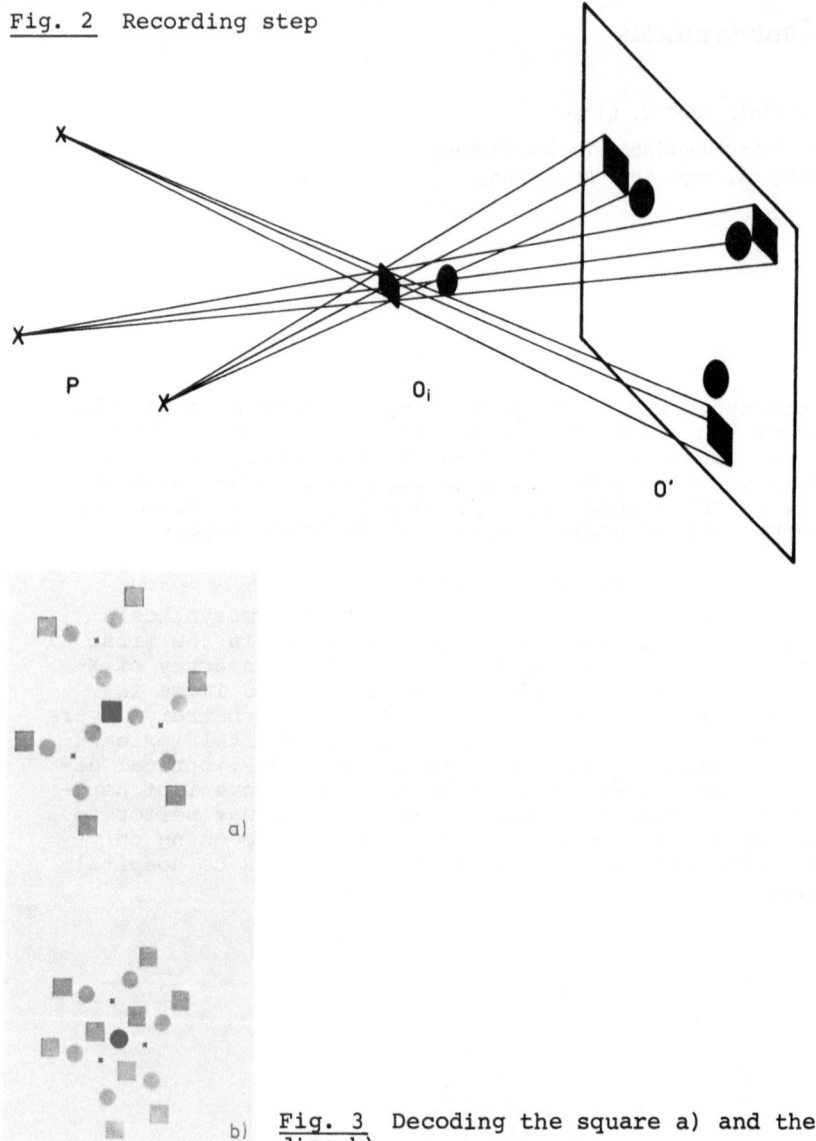

Fig. 3 Decoding the square a) and the disc b)

The coding-decoding procedure is sketched in Figs. 2 and 3.
The array of X-ray tubes P - for simplicity only three are
shown - is projecting the object O, consisting of layers O_i -
only two in Fig. 2 are shown -, resulting in the coded image O'.
Neglecting the size of the focal spots of the X-ray tubes the
point-spread functions P_i corresponding to layers O_i are given
by
$$P_i = \sum_{l=1}^{n} \delta (\vec{x} - \lambda_i \vec{x}_1) \qquad (1)$$

where δ is the Dirac-function, $\vec{x}_1$ the coordinates of the focal spots and λ_i is a scaling factor characterizing the depth of layer O_i. This scaling transformation of details in different depth due to the shadow projection geometry is used for selecting layers in the decoding step.

The coded image O'

$$O' = \Sigma \, O_i \ast P_i \qquad (2)$$

in the decoding step is correlated with the point spread function P_1 for reconstructing layer O_1:

$$O'' = \Sigma_i \, O_i \ast P_i \otimes P_1 = O_1 \ast P_1 \otimes P_1 + \Sigma_{i \neq 1} \, O_i \ast P_i \otimes P_1. \qquad (3)$$

In Fig. 3a this correlation has been performed for the layer of the square. Correlation with a point-distribution P means multiplying the coded image O' according to the number of tubes, shifting these images according to the coordinates $\vec{x}_1$ and summing up the pictures. The result is a three-fold overlapping of the square in the middle of Fig. 3a surrounded by sub-images of the square as well as of the disc. This noise in the reconstruction is due to the fact that in (3) the auto-correlation $P_i \otimes P_1$ is only an approximation to a δ-function, and the cross-correlations $P_i \otimes P_1$, $i \neq 1$, are only approximately cancelling. Nevertheless using n X-ray sources one obtains a ratio of n/1 of reconstructed images to each of the sub-images. That the sub-images only have intensity 1 is due to the very special distribution P of the focal spots. This so-called non-redundant point distribution [3] (NRPD) can be calculated by iterative computer programs taking into account that all possible pairs of coordinate differences within P must not occur more than once.

The layer of the disc easily is then decoded by a scale transformed shifting according to the scale λ_k of the corresponding layer O_k (Fig. 3b). In this way arbitrary layers of the object can be continuously decoded by a continuous scale transformation. This simple selecting of layers is one of the reasons optics is so well adapted for implementing coded aperture imaging algorithms.

2. The Hardware System

In Fig. 4 the arrangement of 24 X-ray tubes is shown, which simultaneously can be flashed in the range of milliseconds. Thus, also moving objects like the pulsating heart of fast flowing contrast media can be imaged from different views. The distribution in which the tubes are arranged is an NRPD shown in Fig. 5a. Its autocorrelation consisting of a middle peak of intensity 24 - responsible for the reconstructed image - surrounded by n(n-1) subpeaks of intensity 1 - causing the noise in the decoded layer, is shown in Fig. 5b. If all n(n-1) side peaks would contribute to the reconstructed layer, the signal to noise ratio would be

$$\frac{n}{n(n-1)} \; < \; 1!$$

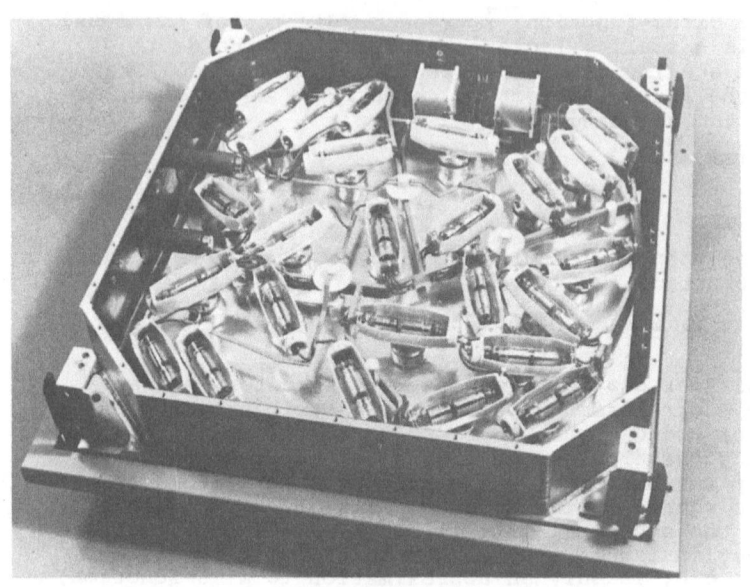

Fig. 4   Array of 24 X-ray tubes

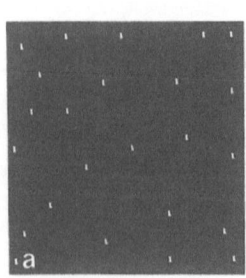

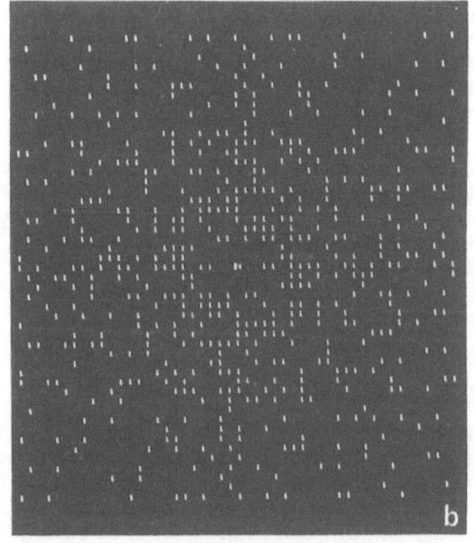

Fig.5
NRPD for n = 24 (a)
and its autocorrel-
ation (b)

This means that the overlapping in the primary coded image of
different views should be kept as low as possible. Each side
peak characterizes an overlapping of the two perspective images.

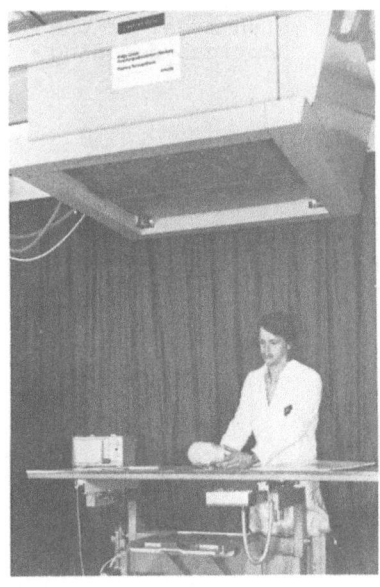

Fig. 6 X-ray equipment for flashing tomosynthesis

The total X-ray equipment is shown in Fig. 6. The holes where the X-rays are penetrating the container are clearly visible.

The decoding procedure of multiplying, shifting and adding the coded image is simple enough for an optical set-up the principle of which is sketched in Fig. 7.

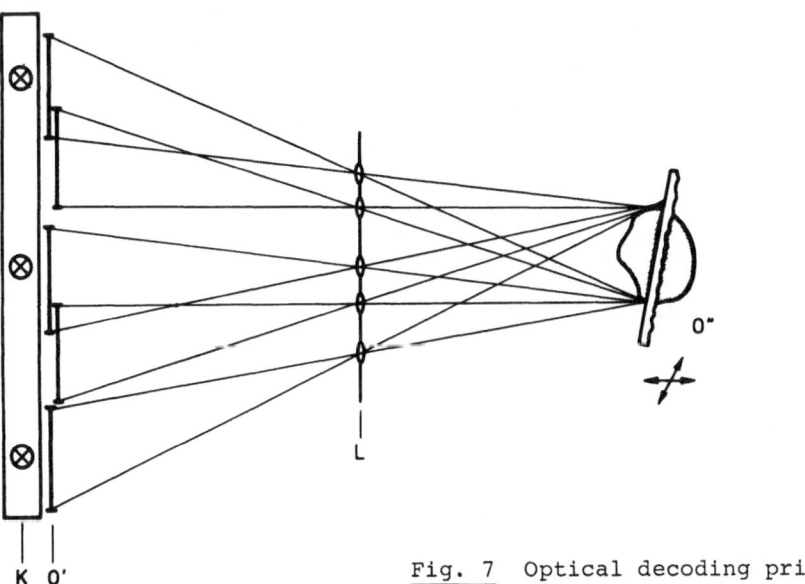

Fig. 7 Optical decoding principle

The coded image O' is placed in front of a light box K and imaged n-times by the lenses L which are mounted according to the NRPD used for arranging the X-ray tubes. Because of this multi-projection a real image O" of the object is reconstructed in the approximation mentioned above. If now a screen is inserted in any arbitrary direction within the reconstruction volume the

97

Fig. 8
Lab-model of the
optical decoding

corresponding tomogram is "cut out" of the object. This offers
the possibility of continuously "going through the object". An
optical device realizing this principle is shown in Fig. 8.
This new optical decoding scheme is superior to the formerly
proposed holographic method because of the possibility of displaying now also tilted layers, or even arbitrary curved surfaces of the object.

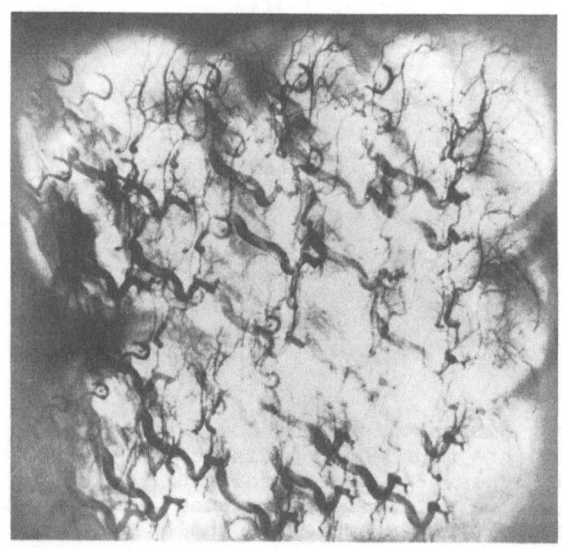

Fig. 9
Coding a brain slice

3. Results

In Fig. 9 a typical coded image is shown. The object was a slice of a human head phantom the vessels of which were prepared with contrast medium. Fig. 10a shows two tomograms reconstructed from Fig. 9 using the device of Fig. 8. The object diameter was 90 mm. The artifacts are caused by the side peaks of the correlation functions. Fig. 10b presents a tomogram, where the object field was 120 mm within another part of the same head phantom. Fig. 11 presents the coded image (11a) of a skull phantom imaged in lateral position and two reconstructed tomograms showing the sella turcica (11b) and the joint of the jaw bone (11c).

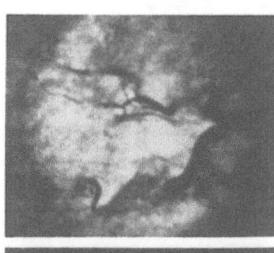
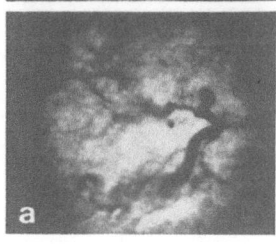

Fig. 10 Reconstructed tomograms of the brain slice

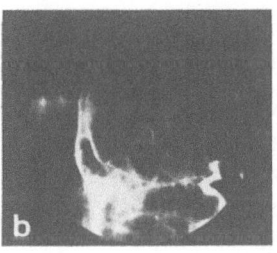

4. Future Aspects

Main task of the future is to reduce the artifacts in the reconstructed tomograms. This means that the side peaks of the correlations (Fig.5b) either have to be reduced in intensity or in number. Both directions were followed in the past years by several authors [4]. A result achieved by one of those methods is shown in Fig.12. It shows a tomogram of the head slice. Comparing it with Fig.10a, the absence of artifacts is quite obviously. Figure 12 was obtained by coherent inverse filtering. Although the method gives very good decoded images, at the time being seems to be a little bit too tedious for installing in a hospital.

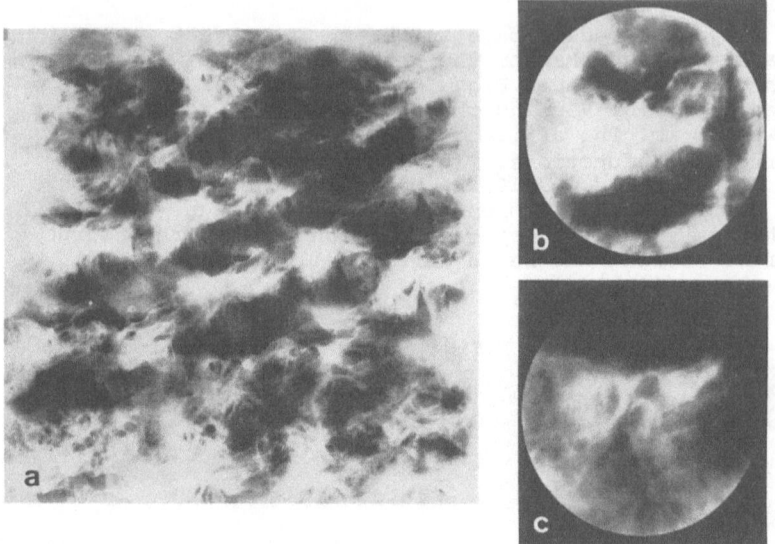

Fig. 11 Coded image of a skull (a) and two decoded tomograms (b), (c)

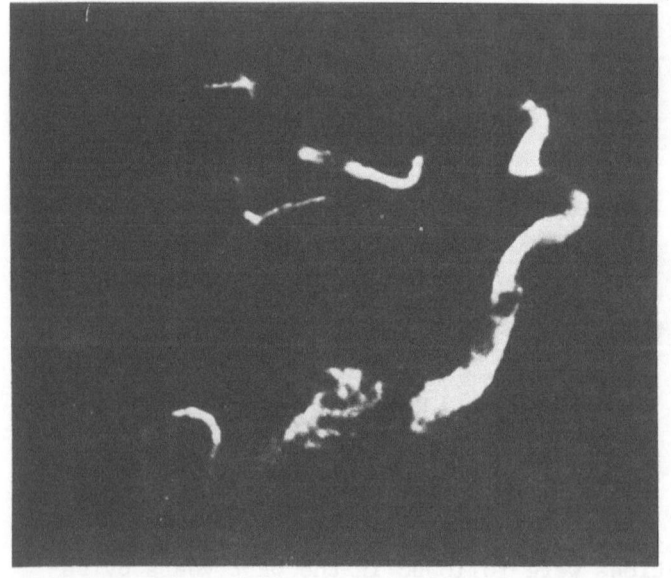

Fig. 12
Tomogram obtained by inverse filtering

This work has been supported by the medical and technological program of the Federal Department of Research and Technology of the FRG within the project LMB. The authors alone are responsible for the contents.

References

   Weiss, H., Klotz, E., Linde, R.
   Optics Laser Technol. p. 117 (1975)
2 Klotz, E., Weiss, H.
   Optics Comm. 11, 360 (1979)
3 Golay, M.J.E.
   J. Opt. Soc. Am. 61, 272 (1978)
4 Weiss, H.: *Holography in Medicine and Biology*, Springer Series
   in Optical Sciences, Vol.18, ed. by G. von Bally (Springer,
   Berlin, Heidelberg, New York 1979)

Holographic Methods of 3-D Representation from a Number of Plane Images for Medical X-Ray Diagnostics

I.R. Utyamishev and R.I. Utyamishev

All Union Research Institute for Medical Engineering
of Public Health Ministry, Moscow, USSR

1. Introduction

The problem of 3-D visualization of biomedical objects attracts
an increasing attention of medical specialists, especially for
studies of the size, localization, and shape of pathological
structures. The analysis of these parameters from plane images
is rather difficult and requires some experience and high quali-
fication of the investigator. In many cases, especially when
using a penetrating type of radiation an inadequate image interpretation is possible. Stereoscopic and holographic techniques
are advanced methods of volume representation. The traditional
stereoscopy is a sufficiently simple and cheap method but does
not produce full spatial visual illusion because the observation
area is limited and parallaxes do not change when a viewer moves
his head. Multiperspective lenticular and integral photographic
systems reproduce scenes with changeable parallax but the corresponding production technology is complicated and not worked
out enough.

Direct holographic recording can be applied only in the visible
optical range for external organs and body surfaces or easily
accessible cavities (stomatology, cosmetology, pathoanatomy,
gynaecology). The possibilities of registration of internal structures are restricted for the following reasons:

1. lack of coherent radiation sources and corresponding optical elements in X-ray or gamma-ray ranges;

2. need of recording materials with extremely high resolution;

3. difficulties arising in the reconstruction process owing
to the great differences between recording and reconstructing
wavelengths.

Moreover there are essential requirements of dosimetry and
mechanical object stability.

In this situation for angiographic, gastrologic, electron
microscopic applications quasiholographic reproducing methods
combining the advantages of stereoscopy and holography are more
convenient and preferable.

2. The Basic Principles of Holographic Stereosynthesis

The hologram is known as a universal information carrier which provides the possibility to form the wavefront coming from an object which is located at any point in space. Contrary to classical photography it is possible to record many images sequentially and reproduce them simultaneously without noticable mutual disturbances what is practically impossible by other means. This striking property can be successively employed for 3-D-image-synthesis from a series of plane photographs taken in incoherent radiation.

The whole process of stereosynthesis may be divided in three main stages of initial polypositional survey, holographic recording and reconstruction. At first many radiographs will be produced from different view points. There are few kinds of radiation source and image registration system motion trajectories such as linear, circle, cicloidal, etc. The most preferable for stereographic purposes are linear tomographic and circular roentenographic arrangements. In case of circular motion (see Fig.1a) exposures must be done in constant angular increments of emitter and receiver rotation. In longitudinal tomography as shown in Fig.1b X-ray tube 1 and fluorescent screen 3 are moved in opposite, mutually parallel directions. The pictures are recorded in every linear tube displacement step d. The corresponding shift of the imaging system 3 is $d_2 = d_1 \cdot \rho / f$, where ρ is the distance between scanned section plane and intensifying screen, f - tube focal length, d_1 - displacement of the tube.

The second operation is the synthesis of the shape image directly from a number of radiographs produced at the previous stage. The holographic recording arrangement must be in absolute conformity with the geometry of the radiographic equipment. All existing stereoholographic methods can be classified by two main groups: multiplexed [1] and composite [2] stereograms. The strictly spatial image separation is a general requirement for qualitative visual perception. For the first type of stereograms a sequential holographic superimposing of many viewpoint images having respectively different object beam orientation is produced. Image plane holograms are usually recorded [3]. Reconstructed perspectives have nonoverlapping exit pupils in space. When a composite stereogram is synthesized, every projection is recorded on a separate area of a holographic plate through a narrow optical slit as a vertical strip, and as a result each of them reproduces only a single image.

Angular separation arrangements are usually used for a circular motion radiographic apparatus and slit schemes are preferable for a linear tomographic trajectory. But in some cases, for cylindric shaped roll film stereograms with an observation angle of 360° circular photographic survey and slit view separation can be used [4].

The reconstruction process can be realized by means of a monochromatic or white light point source depending on the recording arrangement (transmission, Fourier, rainbow, image plane or reflection hologram types).

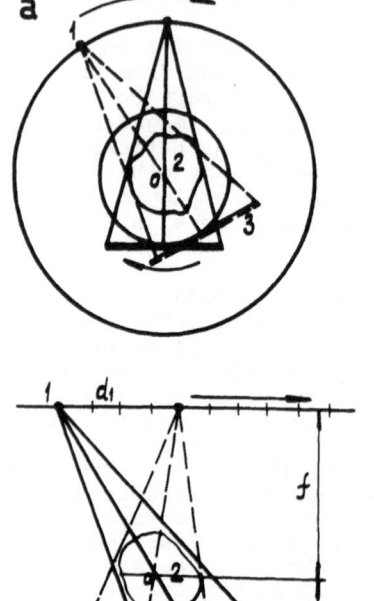

Fig.1a,b. Multiperspective radiographic arrangements (a) circular geometry, (b) linear tomography 1 - X-ray source; 2 - biomedical object; 3 - imaging system

3. Stereographic Image Quality Analysis

The quality of multiple stereographic pictures, their plasticity and resolution depends on the discretization degree or the number of plane images and the value of angular intervals between them. The elementary holograms size choice is essential for the slit separation method. When their sizes are too small the diffraction effects degrading the resolution are considerable. On the other hand when their apertures are sufficiently large the 3-D scene to be reproduced is continuous and stable. The transition of an object image point when the observer changes the viewpoint looking through the neighbour element must be comparable with the optical system resolution Δr [5].

$$\Delta r = 1,22 \frac{\lambda R^2}{d\delta}$$

where δ is the diameter of human eye pupil; d - stereogram element size (slit width); R - distance from image plane to viewer; λ - light wavelength.

As the number of perspectives is finite, dead and overlapping areas as well as distortion appear in a composite hologram owing to the fact that a slit stereogram is watched through all the elements S_i, representing different angles of view which have their own viewing centers E_i (see Fig.2).

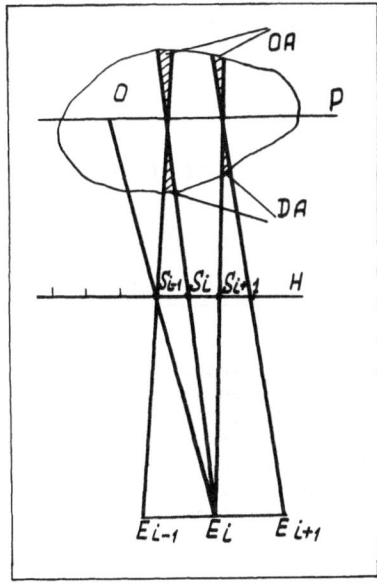

Fig.2 Discretization effects of composite stereogram O-object, P-image plane, H-hologram, S_i-i-th stereogram element, E_i-viewpoint corresponding to i-th element, DA-dead areas, OA-overlapping areas

The first phenomenon is practically negligible for visual perception when the discretization degree is sufficiently high and the second one is increasing for deeper planes and peripheral stereogram elements. Superimposed holograms may have abrupt jumps between neighbour images because of free spaces or inversely overlapping ones between them which all together lead to visual discomfort. In addition to the mentioned facts there are many other degrading factors affecting the synthesized hologram quality such as object motion during the radiographic process, the presence of X-ray apparatus and multiplexing system rotation excentricity, nonconsistency of linear and angular relationships in radiographic and stereosynthesis stages.

4. Radiologic Dosimetry Problems and Some Technical Aspects of Synthesis

As mentioned above for 3-D quasiholographic image representation it is necessary to produce multiple X-ray exposures. The number of radiographs can vary from 50 to 1000 for angular quantification degree $1/3 - 1°$ depending on the observation zone angle. The main parameters of roentgenographic image receiving systems are represented in Table I.

Table I. Comparison of characteristics of X-ray imaging systems

System	Intensifying screens		Fluorography			Image Intensifiers (cinematography)	
Type	Tungsten	Ittrium	100mm	70mm	35mm	70mm	35mm
Dose per picture [mR]	1,5	0,3-0,5	12	6-7	0,8-1	0,02	0,02
Resolution [line/mm]	6-7	6-7	6	10	20	15	30

Minimal patient radiation energy can be provided when intensifiers are used. Thus, fine photographic quality can be achieved for large format X-ray pictures with high resolution and wide density dynamic range. Cardiologic and angiographic researches, however, need the recording process to be done in possible short time. In this situation it is logical to adapt modern tomographic equipment, namely linear tomographs and axial computer scanners with high speed mechanical units operating in pulsed rapid cinematography regime.

Excellent results can be expected in microroentgenography and histology for pathology and internal organs specimen demonstration. There are no restrictions of dosimetry and it is not necessary to employ expensive and bulky installations. Low power X-ray tubes with very small focal spots can be used. That makes it possible to record high resolution images with a good soft tissue contrast transfer.

The metrology and optimization of holographic photosensitive material parameters is one of the central problems in stereosynthesis. The constant image brightness from any point of view, the invariability of hologram geometrical parameters and their conformity with the radiographic arrangement are the main conditions for a qualitative representation.

As known after chemical processing photographic emulsions undergo shrinkage which affects negatively the stereoscopic reproduction by violating the angular relationships and decreasing the image intensity. Calculations and experimental studies show that the deviation of the reconstructed object beam direction can reach $5-10^\circ$. For multiplexed stereograms image brightness decreases when many holographic recordings are superimposed with the number of sequential exposures rising. Practically the superposition number has to be limited to 30.

5. Experimental Study

Stereoholographic arrangements as well as slit and angular view separation, image plane, reflection and transmission type recordings were investigated. An experimental set-up was developed (see Fig.3) to realize different methods.

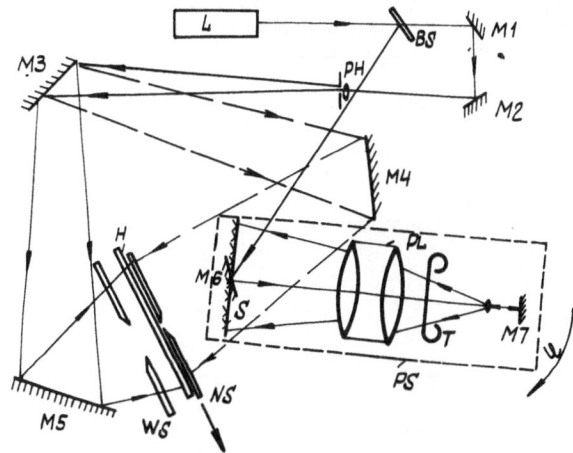

Fig.3 Experimental optical arrangement, L-laser, BS-beam-splitter, M-mirrors, S-screen, H-hologram plate, PH-pinhole, PL-projection lens, WS-wide slit, NS-narrow slit, PS-projection system, T-X-ray photographic transparencies

The projection system PS can be rotated around a fixed axis laying in the screen plane S. The perspective radiographs are projected on the diffusely transparent screen S. The narrow optical slit S in the vicinity of the holographic plate H is used for composite stereogram recording. It is not necessary to change the orientation of the object beam for a linear tomographic scheme because the image plane S and the plate H should be parallel. When focused image holograms are recorded the screen S and the slit NS must be removed, and the projection system PS must be refocused to the H plane.

The following modifications of the optical arrangement for reflection hologram recording were made (see Fig.4):

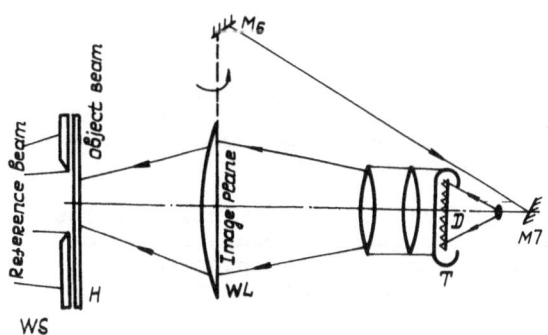

Fig.4 Improved arrangement for reflection hologram recording

The screen S was replaced by a large aperture plano-convex lens WL and a diffusor D was introduced behind the transparency T. The diffusor D is used to eliminate the bright image of the point source of light behind the slide and to increase the effective observation area of the different radiographs producing the hologram. The convergent lens WL makes the object beam angular diagram narrow and does not influence the image plane position. The exposing area of the object beam is smaller than the holographic plane H dimensions. Thus, it is possible to reduce the number of superpositions per surface unit by means of a movable mask WS in the reference beam path.

Many 3-D images were obtained from 2-D radiographs or photographs of different biomedical objects, such as X-ray phantoms or contrasted tissue samples. Some photographs of holographic reconstructions are presented in Fig.5.

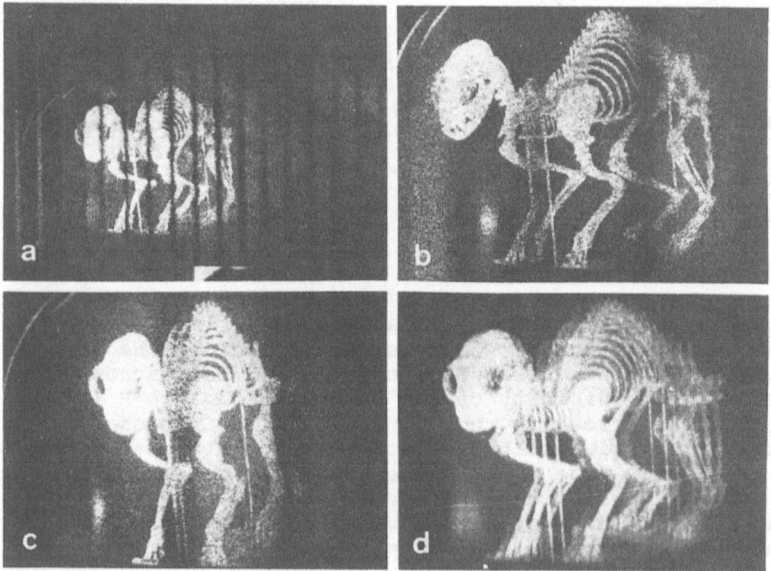

Fig. 5a-d Photographs of reconstructed images (the skeleton of a cat) from a slit stereogram, (a) - general view, (b) and (c) - views at different angles of view, (d) - image disintegration effect

We must take into account that the characteristics of the reconstructing light source must be identical with those of the reference one. Disregard of this requirement can lead to distortions, displacements or disintegration of volume images.

Theoretical estimations and preliminary experimental results show the principle possibility to use quasiholographic methods for medical diagnostics, but to be sure of their practical realization many further studies have to be carried out.

References

1. J.D. Redman: SPIE Seminar-in-Depth Holography, p.161,1968
2. H.H. Chau: Opt. Commun. *4* (1), 1 (1971)
3. N.D. Haig: Appl. Opt. *12* (2), 419 (1973)
4. T.H. Jeong, H. Snyder: Society for Information Display International Symposium Digest of Technical Paper V, 1974
5. T. Yatagai: Opt. Commun. *15* (11), 272 (1976)

Three Dimensional Holographic Synthesis (T.H.I.S.) of X-Ray Pictures

M. Grosmann, P. Meyrueis, and J. Fontaine

Groupe de Recherche et d'Expérimentation en Photonique Appliquée
Laboratoire de Spectroscopie et d'Optique du Corps Solide
(associé au C.N.R.S. no 232), Université Louis Pasteur
3, rue d l'Université, F-67084 Strasbourg Cedex, France

1. Fundamentals of Holographic Synthesis of 3-D Images

The pioneering work for reconstruction of multi-parallax images by holography has been done by DE BITTETO. His purpose was reducing information quantity for teletransmission of holograms. He continuously juxtapositioned many small holograms. Each of them was reduced to a narrow vertical band containing all the information of the image. He demonstrated feasibility on some plates and determined the optimal width w for a slit-former at 1 mm approximately, and the optimal height h avoiding discontinuity of the image, at

$$R = \frac{d_1}{d_1 + d_2} = \frac{w}{h}$$

where d_1 is the image - hologram distance and d_2 the hologram - observer distance.

The principle of the system is shown in Fig.1.

Each image element is recorded by classical holography as follows: through the vertical slit a part of the plate is exposed to the reference beam and to the object beam. Other parts of the plate are shadowed. Between each exposure the object is slightly turned and the slit is translated by one step. All the plate is therefore exposed strip by strip.

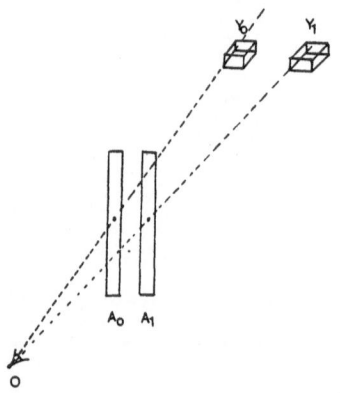

Fig. 1 Principle of holographic reconstruction of multi-parallax images, A_0, A_1 - strip-holograms; Y_0, Y_1 - corresponding reconstructed images; 0 - observer

In the resulting band-hologram the bottom and the top of two contiguous strips contain the information of the adjacent bands.

In Fig. 1, A_0 and A_1 are the strip - holograms recorded on the plate, Y_0 and Y_1 are the corresponding reconstructed images. The observer at 0 sees only the part between Y_1 max. and Y_1 min. of the image corresponding to A_1.

If $d_1 = d_2$ it is clear that 50% of the information on one strip is also contained in the adjoining strips. If the strips are 1 mm wide, this is not disturbing.

There is also a distortion of the wavefront shape in the vertical direction.

This distortion can be corrected by using a cylindrical wavefront.

DE BITTETO's idea was applied by POLE (of I.B.M.) to the treatment of computer generated holograms. He used the sequentialised parallax method (strips) to computer constructed bidimensional images which were looked at through the "window" of the hologram.

DE BITTETO thought that this could also be used for bidimensional images made by classical photography. In this case, the recording can be done without a laser. He therefore used a simple photographic camera to record the images. Lateron in the laboratory he recorded the hologram by projecting the photographic pictures on a ground glass screen before a photographic plate through a slit, and adding a reference beam.

To go from one strip-hologram to the next he had only to translate the strip by 1 mm while simultaneously, translating the photographic film from one picture to the next. He thus recorded what was the first holographic film - a toy car race (70 seconds).

Lateron S. Benton of Polaroid Laboratories developed the rainbow hologram technique, recording a hologram through a slit or with a cylindrical lens.

L. CROSS put these two ideas together to develop the Multiplex Hologram technique. He used a 35 mm movie camera instead of the simple camera of DE BITTETO but kept the principle of the rotating stand for the object. But instead of a ground glass screen he used cylindrical lenses forming the image directly on the screen (a little behind it to avoid aberrations). Since classical lenses of large aperture are very expensive and rather noisy in coherent light he also developed large liquid lenses using mineral oil.

2. The Three-Dimensional Image Synthesizer

2.1 Purpose of the System

This system is designed for synthesizing a three-dimensional picture from a collection of bi-dimensional pictures, for medical applications.

2.2 Principle of the System

The system uses the property of a hologram to give the illusion by looking at the holographic picture of an object as if we were looking at the real object through a window.

If we juxtapose a collection of these "holographic windows", each of them having registered a bi-dimensional picture with a continuous change of point of view, an observer's eye looking through this collection of holograms will perceive the image of the object with an "angular scanning". Additionally, there is the stereoscopic effect: each eye looks through a different hologram with a different point of view of the object. We obtain a strong three dimensional perception.

This property is the direct result of the possibility offered by holography to superpose sequentially spaces without interactions between them. We have especially worked on the three following points:

a) Shooting of the pictures

Very drastic conditions are demanded for direct recording of a holographic picture : laser, darkness, absence of vibrations, sophisticated optical equipment, etc.. We can record almost the same quantity of information with much less trouble. The main simplification is dividing the process into two steps:

- recording of the primary picture by an ordinary photographic process with some care in the lighting to avoid three dimensional shadows.

- transforming pictures into a composite hologram. The horizontal parallax is lost in the process but this is not very important usually.

b) Shape of the hologram

The usual shape of a hologram is rectangular with a ratio 1 x 2 or 1 x 4 for the sides. Our endeavour was to improve the quality of the 3-D perception. If we use too wide a hologram. we are going to have scintillating effects and discontinuities. For removing this, it is necessary to reduce the spacing and the width of the hologram, the width of the slit being given by the size of the eye aperture and the average distance of viewing. We have experimentally determined a width of the slit of 1 mm separated by a 0.6 mm space. The THISGRAPHY is thus composed of a collection of strip-holograms on a high resolution film. The quality of a THISGRAPHY is just limited by the quality of the available commercial films.

112

c) Holographic reconstruction

For display a classical hologram needs a coherent source. Our system being designed for use in medical surroundings must work with a simple white light system. Our solution was to transform the classical hologram during the recording into a rainbow hologram by a lens-system.

3. The Technical Equipment

3.1 Principle of the Process

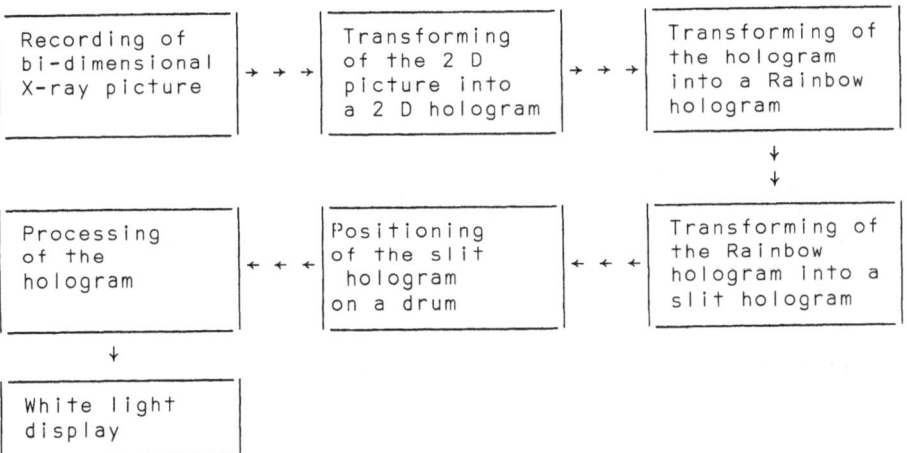

3.2 Components

a) Input of the bi-dimensional data

A special X-ray head is connected to a micro-computer for turning around the objects in well defined steps. Each X-ray picture is then processed by a video movie converter and recorded with a special 35 mm camera (modified for stable recording).

b) Holographic processing (Fig. 2)

A filtered and collimated laser beam lights a black and white picture of the 35 mm film. The picture is then processed via a set of crossed cylindrical lenses. The purpose of these lenses is to create the rainbow effect and the slit effect. These lenses are made of plastic sheets, containing mineral oil, or of plastics. The design has been studied to avoid aberrations at the He-Ne wavelength. In the case of the liquid lenses, the shapes of the wooden frame have been designed with a computer. The curves have been reproduced in the cut of the wooden material. In the case of the plastic lenses a computer controlled cut has been tried. But we have only obtained good results with cylindrical shapes.

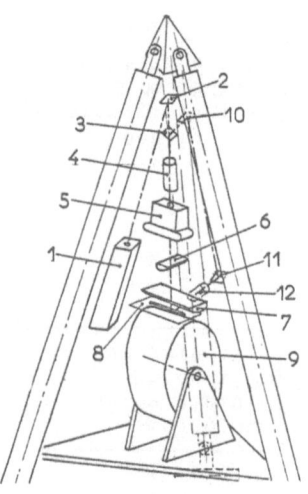

Fig. 2 General view of the T.H.I.S.: *1* laser; *2* mirror; *3* beam splitter; *4* beam expander and shutter; *5* film holder with feed back; *6* lenses for rainbow effect; *7* slit effect lenses; *8* slit holder; *9* positioning drum with vacuum held hologram; *10, 11* mirrors for reference beam; *12* beam expander and shutter for reference beam

The reference beam angle is 50° for the best diffraction efficiency at usual vision angles. Our lenses produce practically no noise and distortion on the strip-hologram.

c) *Recording of the final hologram*

For recording a coherent global strip-hologram picture, it is necessary to position the film with high accuracy. The translation between each shooting must be done with at least 1/10 mm precision, without any vibration.

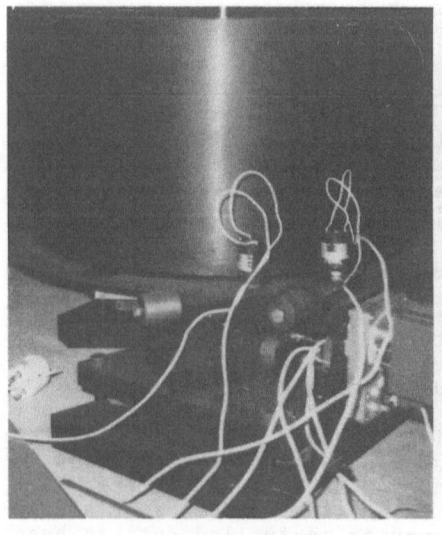

Fig. 3
Drums for positioning the holograms

Fig. 4
Film positioner

Fig. 5
Horizontal liquid lens

Fig. 6
Twin liquid lens-system

This is obtained holding the film by vacuum on a drum with a diameter of 50 cm. The positioning of a holographic image is automatically connected with the positioning of the movie picture. Thus, every movie picture corresponds to a strip-hologram. A photosensitive device measures the intensity of the different beams. A microprocessor adjusts automatically the different ratios of light intensities. The system is completely automatized. The only manual operation needed is the loading and the unloading of the films.

The processing of the film is common, the exposure time is about 1 second per strip with a 100 milliwatt laser. We need about half an hour for a 360° THISGRAPHY.

d) Structure of the system

The system is installed in a composite constructed of wood, sand, concrete and steel. Its weight is about 6 tons. It is necessary to instal the system in a steady room and to avoid work of compressors, engines etc. during the recording procedures.

The components are linked to each other through a rigid frame mounted on shock absorbers, a vibration control system shuts off the process when the vibration exceeds the maximum level permissible.

e) Display in white light

Our white light source is a halogen car headlamp that provides maximum spatial coherence, a vertical filament for improving the rainbow effect, and, in addition, low cost as well as easy availability.

The THISGRAPHY is positioned in a transparent plastic holder above the lamp at an angle of 50°. The viewing angle is around 50°. The observer's eye is in the image of the slit produced by the system of lenses.

The image of this slit is enlarged by the length of the lamp filament. THISGRAPHY is curved to compensate for horizontal aberrations and to improve 3-D effects.

References

1. D.J. de Bitteto, "Bandwidth Reduction of Hologram Transmission System by Elimination of Vertical Parallax", Appl. Phys. Letters, Vol. 12 - Nb. 5, (March 1968).
2. S.A. Benton, J. Opt. Soc. Am. 59, 1545 (1969).
3. L. Cross, private communication (1975).
4. O. Koshi, Three Dimensional Imaging Techniques, Acad. Press (1976).
5. P. Meyrueis, M. Grosmann, French Patent (1979)

Tomogram – Reconstruction by Holographic Methods

H. Platzer and H. Glünder

Institut für Nachrichtentechnik, Technische Universität München
D-8000 München, Fed. Rep. of Germany

1. Fundamentals of X-Ray Tomography

1.1 Introduction

The aim of X-ray tomography is to reconstruct the two or three
dimensional density function of an object from projections.

The elementary measurement is the chain product of absorption
values in the path of a ray emitted by one point of the X-ray
source and recorded at another point in a detector plane by
X-ray film or a scintillation counter.

Switching to the logarithms of the absorption values, this
elementary measurement turns out to be the formation of the
line-integral between the emitting and the detecting point. All
the values of the line-integrals together give the coefficients
of a system of linear equations which is to be solved for the
density function of the object under consideration.

1.2 Noise

This reconstruction of the object means essentially, that every
absorption value is obtained as the difference of two nearly
equal linear combinations of the measurements.

These measurements are subject to quantum noise, so that we
always obtain m counts or silver grains with an uncertainty of
$\sqrt{m}$ counts respect. silver grains. The total number of photons
to be recorded or the area of the recording-film depends on
image quality parameters of the reconstructed object. In case
of a reconstructed cross-section with a linear resolution of n
and an amplitude resolution of 1/a the number N of photons to
be counted or photographically recorded is:

$$N = c \cdot n^3 \cdot a^2;$$

where n is the number of linear independent values along the
diameter of the reconstructed image; 1/a is the relative error
of the reconstructed amplitude; c is a constant in the order
of magnitude of 1.

When we try to record the projection on X-ray film, the required area of film has the order of magnitude of the size of
one developed silver grain times the number of photons given by
the above equation.

If we choose a spatial resolution of 1000 x 1000 pixels and an amplitude definition of half a Hounsfield-unit (one part in 1000), we require:

$$(10^3)^3 \cdot (2 \cdot 10^3)^2 = 4 \cdot 10^{15} \quad [\text{counts}] \; ;$$

Having a silver grain of the size of one micron we would need:
$4 \cdot 10^{15} \cdot 10^{-12} \; [m^2] = 4 \cdot 10^3 \; m^2$ of film for one tomographic recording.

For another set of image parameters:
Linear resolution 256, a 5 bit amplitude resolution, one finds that already a piece of X-ray film of (10 x 10)cm^2 is sufficient under the same assumptions. So, there is a region of parameters, for which it is worth looking for photographic recording and methods of analog reconstruction.

1.3 Superpositions of Projections

It must be noticed that reconstruction by linear operations is only possible if the quantities under consideration (the logarithms of densities) are linearly combined. This holds for the well-known computer aided tomography (CAT), for the sinogram recording and for related methods. But it does not hold in general for coded-source imaging where the several signals from elementary measurements overlap at one point in the detector plane, and where they are added before a logarithmic transformation could take place. This is one of the fundamental drawbacks of most coded source imaging techniques. Another is discussed in the next section.

1.4 Depth Resolution in Coded-Source Imaging

Another drawback of coded-source imaging is that the achievable set of measurements is incomplete, so that the reconstructed image carries certain interdependencies, i.e. blurring which results from the finitness of the source configuration and which, unfortunately, cannot be corrected in any way, because of the theoretical reasons given below. In spite of this, many attempts have been made, to achieve depth resolution by coded-source imaging. Such source structures may be carefully constructed point-arrays, among others [1]. The system philosophy is, to focus the various depth layers by filtering the superposition of the coded images of all layers with the proper filtering function. The aim is to achieve a delta-function-like system response for the layer under consideration and a negligible contribution of the other layers.

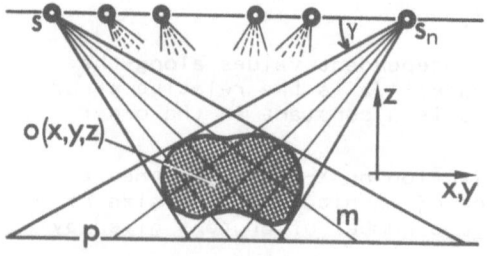

Fig.1 Imaging of an object o onto a detector plane p by a coded X-ray source $s_1 \cdots s_n$

Elementary measurements are shown as straight lines m

Let us consider for example a 3-D body whose x,y-layers should be reconstructed for various values of the depth-coordinate z. Fig.1 shows the system configuration. In the f_x, f_y, f_z-Fourier-space, the regions of no signal can be found by using the relation between a projection in the space domain and its corresponding representation in the spatial frequency domain, known as the "central slice theorem" [2]. (Fig.2)

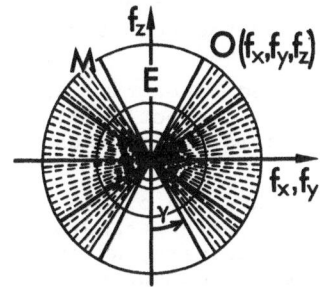

Fig.2 Fourier domain representation of elementary measurements M with the empty cone E

Thus, the range of spatial frequencies which remains totally unfilled can easily be seen to depend mainly on the aperture of the imaging system and also on the size of the object. It can be seen, that the cone around the f_z-axis has the value zero, with all finite coded sources. So it is useless to look for algorithms to improve the depth resolution for spatial frequencies f_x, f_y inside this empty cone [2].

It can also be seen that there exists a depth resolution proportional to the spatial object-frequency and the aperture of the illumination. Apriori knowledge like that the object is positive valued cannot be used within the domain of methods this paper is dealing with and which is characterized by linear filtering and coordinate transformation.
Similar considerations hold for coded aperture imaging.

2. A Survey of the Present Analog Transaxial-Tomography Reconstruction Methods

In CAT three reconstruction methods are used to obtain the 2-D image of an object-slice from 1-D projections:

Recursive	(2.1)
Correction by convolution (space domain) + back-projection	(2.2)
Synthesis in the frequency domain + correction-filtering (rho)	(2.3)

2.1 Recursive methods are well suited for implementations on digital computers and are not suitable for analog techniques.

2.2 Convolution + back-projection is presently the most widely used technique of CAT, because of the development of fast parallel computers (pipeline- and array-processors) for convolution operations and coordinate-transformations.

2.2.1 EDHOLM et al. [3] achieved good reconstructions by the corresponding optical analog technique (incoherent), using as for most of the optical methods the sinogram as input-data.

2.2.2 Some amazing results from the Optical Sciences Center, Tucson [4] were obtained by techniques still using the convolution for correction, but reconstructing the object pointwise. The drum- and loop-processor performs the convolution optically in an incoherent mode and after opto- electric conversion the reconstructed image is synthesized on a CRT-display.

2.2.3 The commercially available Simtomix-Analog-Tomograph [5] has a similar concept. It works on TV-basis and the convolution is also carried out optically (incoherent). So the reconstruction has TV-quality.

2.2.4 Another method starts with the so called layergram (backprojection) followed by a 2-D coherent-optical correction-filtering [6]. This technique uses the Fourier-domain, and thus already leads to point 2.3.

2.3 The synthesis in the frequency domain makes use of the central-slice-theorem. A great number of 1-D Fourier-transformations (F.T. s) and one final 2-D F.T. are necessary, which are time-consuming when implemented on a digital computer. Coherent optically, as it is well known, the F.T. appears in the focal-plane of a lens.

1977, STROKE and HALIOUA [7] proposed a holographic method for 3-D reconstructions of specimens in electron-microscopy. It allows one to synthesize in a discrete way the 3-D Fourier-space of the 3-D specimens out of a set of 2-D projections. Paragraph 3. presents our method for 2-D transaxial tomography reconstruction, where the F.T. is built up continuously.

2.4 Some recently developed methods for analog tomogram-reconstruction, which have no digital analogue and which seem to be very promising, should be mentioned before. NISHIMURA et al. [8] proposed a sequential coherent optical filtering, while HANSEN et al. [9] managed to obtain reconstructions fully in parallel by coherent optical inverse-filtering combined with a spatial weighting. HOFER [10] did a system-theory analysis of the latter. Furthermore he showed what the filter function looks like and how the filtering can be implemented in detail. A great advantage of this method is the lack of any mechanically moving parts. A drawback of both methods is the need for modified sinograms which have to be generated by an additional process. Furthermore the reconstruction obtained by the second technique is given in inconvenient coordinates.

3. Reconstruction by Synthesis in the Frequency Domain

The goal for the development of this new technique was to avoid, as far as possible, the complicated mechanical arrangements that are necessary for nearly all of the previously mentioned methods (except for the last-cited). In particular the precise synchronous, mechanical movements of the types listed in table 1 are difficult to achieve and often are sources of errors. For the method which we present here no coupled movements or electronically synchronisations are needed. Furthermore the required adjustments are minimized and may be servo-controlled.

Table 1:

Technique:	Coupled Movement:
back-projection	linear-rotational (rot.)
drum-processor	rot.-rot.-electron beam position (r,φ)/intensity
loop-processor	linear-electron beam position (r,φ)/intensity
sequential-filtering	linear (x) - linear (y)

As it is known from the central-slice-theorem, a 1-D projection of a 2-D object has F.T.-values along a line through the origin of the 2-D Fourier-plane. Thus, if all projections are Fourier-transformed and recorded under the appropriate angle, the whole spectrum of the object is available and the reconstruction may be obtained by rho-filtering and inverse F.T.

The projections which are stored in a sinogram are sequentially Fourier-tranformed. Because these 1-D F.T. s are complex-valued the recording in the frequency domain has to be a hologram. The basic idea is shown in Fig.3.

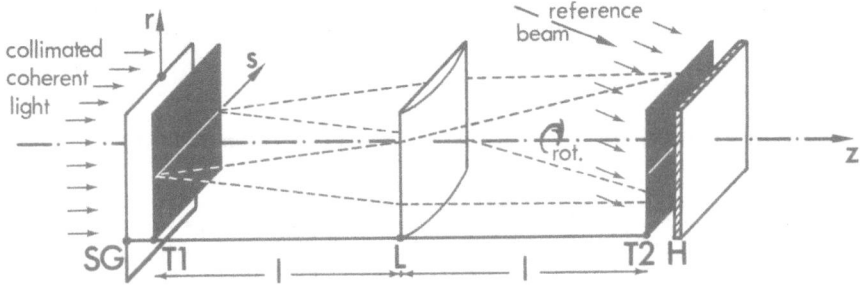

Fig.3 The principle of a processor for frequency domain synthesis

The elements (SG,T1,L,T2) rotate around the z-axis while the sinogram (SG) moves radially in front of the slit (T1) which selects sequentially the continuously recorded projections. These selected projections are Fourier-transformed by the cylindrical lens (L). The F.T. s are successively recorded on the fixed hologram-plate (H). While the whole length of the sinogram passes in front of the slit, the apparatus rotates through 180°. The 1-D slit-filter (T2) in front of the recording-plane serves as a correction-filter. Thus the requirements for the dynamic range of the holographic material are reduced, in contrast to cases where the filtering is done together with the hologram reconstruction. The intensity-transmission characteristic is shown in Fig.4. If the requirements for the dynamic range should be further reduced, filters with an even stronger suppression of the low frequencies might be used during the hologram recording in conjunction with a compensating-filter applied during the hologram reconstruction. Experiments of a sequential, holographic frequency domain synthesis have recently been successfully performed at our institute.

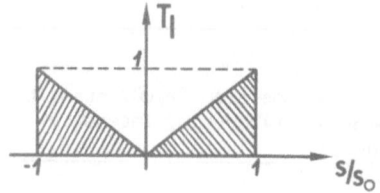

Fig.4 Intensity-transmission versus slit-length of the correction-filter (1-D rho-filter)

The main disadvantage is still the <u>coupled</u> movement (lin.-rot.). A modified sinogram helps one to solve this problem. Instead of recording the sinogram on a linear strip (Fig.5), it is recorded as a halfring (Fig.6).

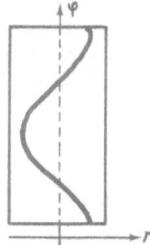

<u>Fig.5</u> Conventional sinogram

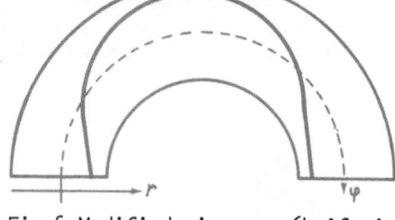

<u>Fig.6</u> Modified sinogram (half-ring)

This special recording may be done directly by X-rays or by geometrically transforming a conventional sinogram, for instance by a cone-mirror. The final arrangement is shown in Fig.7.

Similarly to Fig.3 a slit-mask (T1) scans the sinogram while the cylinder rotates through 180°. The shape of the slit depends on the X-ray beam configuration during the sinogram recording. The mirrors (M1) and (M2) bring the signal-beam S to the optical axis. The lenses (L1) and (L2) together perform the 1-D F.T. The slit-filter (T2) does the necessary correction-filtering. The reference-beam R is collimated by the lens (L2). The sinogram (SG) and the hologram-plate (H), are fixed.

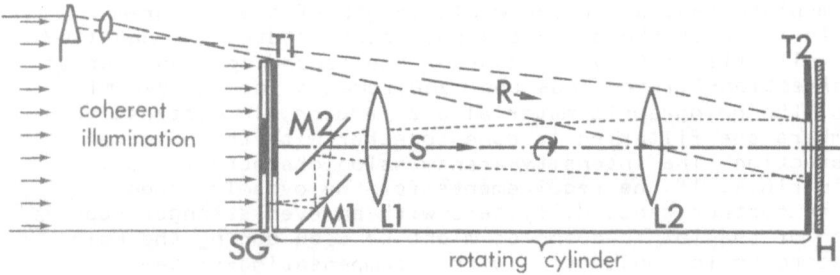

<u>Fig.7</u> Processor for tomosynthesis in the frequency domain without coupled movements

Adjustments

Initial: M2,L1,L2,T2 have to be centerd on the axis of rotation.
For every sinogram: SG has to be positioned, so that T1 and M1
scan in the correct way. If the position of SG is slightly ec-
centric, it would cause a corresponding phase error in the
frequency domain. An interferometric servo-control may help to
maintain the correct position.

On-line processing may be achieved when a thermoplast is used.
for the hologram recording. Such a material allows a hologram
reconstruction immediately after the exposure.

A processor using the principle sketched in Fig.7 is being
built to obtain corresponding experimental results, in addition
to the fundamentally crucial "sequential holographic frequency-
domain synthesis" experiments mentioned above.

4. Acknowledgement

To prepare this paper was especially joyful because of the
many fruitful and friendly discussions the authors had with
Prof.Dr. G.W. Stroke who spent his guest professorship under
a Humboldt-prize at our institute (1978/79).
Thanks also to J. Hofer for very essential criticism.

5. References

1   Weiss, H.; Klotz, E.; Linde, R.; Rabe, G.; Tiemens, U.: "Coded
    Aperture Imaging with X-Rays (Flashing Tomosynthesis)"
    Optica Acta, vol. 24, no. 4, 1977, pp. 305-325
2   Barrett, H.H.; Chiu, M.Y.:"Three-Dimensional Radiographic Imaging"
    Proc. of ICO-11 Conf., Madrid, Spain, Sept. 1978, pp. 135-138
3   Edholm, P.; Hellström, L.G.; Jacobson, B.: "Transverse Tomography
    with Incoherent Optical Reconstruction"
    Phys. Med. Biol., vol. 23, no. 1, Jan. 1978, pp. 90-99
4   Greivenkamp, J.E.; Swindell, W.; Barrett, H.H.; Gordon, S.K.; Gmitro,
    A.F.: "Recent Advances in Uncomputerized Transaxial Tomography I -
    The Loop-Processor" Paper FE 4 - OSA 1978, Annual Meeting, Oct. 78
    "Recent Advances in Uncomputerized Transaxial Tomography II -
    The Drum Processor" Paper FE 5 - OSA 1978, Annual Meeting, Oct. 78
5   Gèluk, R.J. (c/o De Oude Delft):"Verfahren und Vorrichtungen zum
    linearen Filtern von zweidimensionalen Signalen"
    German Patent: DT 2659427 A1, Dec. 29, 1976
6   Peters, T.M.: "Spatial Filtering to Improve Transverse Tomography"
    IEEE Trans. Biomed., vol. BME-21, no. 3, May 1978
7   Stroke, G.W.; Halioua, M.: "Holographic Implementation of our 3-D
    Reconstruction Method for Application in Electron Microscopy of
    non Crystalline Specimens"
    unpublished Report, May 4, 1977; private communication
8   Nishimura, M.; Casasent, D.; Caimi, F.: "Optical Inverse Radon Transf."
    Opt. Com., vol. 24, no. 3, march 1978, pp. 276-280
9   Hansen, E.W.; Goodman, J.W.: "Optical Reconstruction from Projections
    via Circular Harmonic Expansion"
    Opt. Com., vol. 24, no.3, march 1978, pp. 268-272
10 Hofer, J.: "Optical Reconstruction from Projections via Deconvolution"
    Opt. Com., (in print)

Sequential Coherent Optical Reconstruction from Projections Using a Roach Filter

A.M. Landraud and J.J. Clair

Laboratoire d'Optique, tour 13, 3ième étage, Université Pierre et Marie Curie
4, place Jussieu, F-75230 Paris Cedex 05, France

Abstract

A sequential optical technique of reconstruction of a cross-
sectional tomographic plane of a 3-D object, from its projections,
is presented. The data are preprocessed by means of a technique
for coding image information in real time and an analogical inver-
sion of the equation of convolution, involved in the theory, is
achieved with a synthetic holographic filter (R.O.A.C.H.). The
principle of the method is derived from the properties of lineari-
ty of the mathematical operations used in computerized transaxial
tomography. The order of the operations may be inverted and the
treatment be performed sequentially by using one-dimensional
filters or grids in a simple coherent optical processor. The
method may be used to measure a local parameter of a 3-D object,
relatively transparent to electromagnetic energy, such as a
coefficient of attenuation in X.R. Tomography or phase variations
in biological objects. To reconstruct 3-D refractive index or
phase profiles we have designed an optical interferometric set-up
to record the "projections".

1. Introduction

The mathematical operations of back-projection, summation, Fourier
transforms, multiplication in Fourier space, inverse Fourier transforms and convolutions, all computerized in conventional transaxial
tomography, may be performed optically by means of an optical encoding of the information and by using suitable filters and systems
of lenses illuminated by coherent or incoherent light. An optical
system used as an analog computer for data processing can provide
a linear representation of functions with two independent variables
and a bidimensional map of data may be transferred into another one
by transferring all the data simultaneously and instantaneously,
and that permits a powerful and parallel information processing.
We present, in section 2, a special technique for recording projections in the case of biological objects where phase variations
are to be reconstructed. After a brief review, in section 3, of
the principle of optical coherent two-dimensional image processing,
we describe, in section 4, a method to encode optically the information, by grids, on a photographic emulsion. In section 5, we
show how the recorded grids, which are physical back-projections,
are used as input in a coherent optical filtering system involving
a ROACH filter.

2. Recording of projections to measure phase variations in transparent biological 3-D objects or in optic fibers

In classical transaxial tomography [1], a thin layer (the tomographic plane to be reconstructed) of the transparent 3-D object is scanned with a narrow X.R.beam. If θ is the angle between a fixed axis Ox in the tomographic plane and the direction Ox' perpendicular to the straight line joining the source S to the point P in the detector, the measured values, or data to be processed, are the so-called "projections":

$$p_\theta(x') = - \text{Log}(I/I_o) = \int_{-\infty}^{+\infty} k(x',y')dy', \qquad (1)$$

where $k(x',y')$, or $k(M)$, is the X.R. absorption coefficient to be reconstructed from line integrals. I_o is the source intensity reference recorded at point P in the absence of the absorbing object. For each direction θ of scanning, the source S and the detector P are both translated, step by step, and the data are stored. Then the angle θ is incremented and a new set of projections is measured. It has been shown that, mathematically, between the repartition $k(M)$, a simple function $g(M)$ and a so-called "summation image" $h(M)$, exists a relation of convolution:

$$h(M) = k(M) \divideontimes g(M), \qquad (2)$$

where $g(M)=g(r)=1/r$, r being the radial coordinate in the tomographic plane. The function $h(M)$ is proper to the object; it represents, at each point M, the superposition of all the projections, or flows, carried along all the straight lines passing by M.

In the case of biological objects or of optic fibers, the local parameter to be reconstructed is a small variation in the refractive index: $k(M)=\triangle n(M)$. A tomographic plane of the 3-D transparent object is scanned with a narrow laser beam (which does not destroy the biological cells). To collect data we have designed a suitable two-beam interference device [2]. The recorded flows then result from an interferential phenomenon(Fig.1) and the proposed method allows to reconstruct, plane by plane, a map of variations of phase or index in the 3-D object.

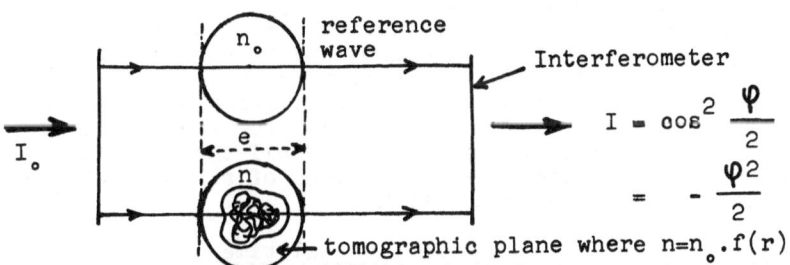

Fig. 1. Schema illustrating the principle of data recording when the parameter to be reconstructed is a refractive index or a phase variation.The recorded energy is proportional to the quadratic phase.

For a given angle θ of scanning, the measured intensity distribution is proportional to:

$$I = I_o \left(\frac{2\pi}{\lambda} \int (n(y') - n_o) dy' \right)^2 \tag{3}$$

By analogy with the above theory we call "projections" the quantities:

$$p_\theta(x') = \frac{\lambda}{2\pi} \sqrt{I} = \int \triangle n(x',y') dy' \tag{4}$$

and similarly we obtain a "summation image" so that:

$$h(M) = \int_0^\pi \int_{-\infty}^{+\infty} \triangle n(x',y') dy' d\theta = \triangle n(M) * g(M), \tag{5}$$

where g(M)= 1/r.

3. Optical coherent 2-D reconstruction from projections

Eq.(5) is formally identical to Eq.(2). The Fourier transform $H(\Omega)$ of h(M) is a simple product of the Fourier transforms $K(\Omega)$ and $G(\Omega)$ of k(M) and g(M) respectively, Ω being a point in the Fourier space:

$$H(\Omega) = K(\Omega) \cdot G(\Omega) \tag{6}$$

with $G(\Omega)=G(\rho)= 1/\rho$, where ρ is the radial coordinate in Fourier space.

A coherent two-dimensional reconstruction of k(M) is possible by first taking the Fourier transform of h(M) and then by performing two successive coherent operations described by:

$$K(\Omega) = H(\Omega) \cdot \rho \qquad \text{and} \qquad k(M) = T.F.^{-1} (K(\Omega)). \tag{7}$$

The optical processing implies an encoding and a recording on a physical support, like a photographic plate, of the information contained in the function h(M). We call "object filter" such a recorded image summation. The object filter is used as input in a double diffraction system with a "ρ" filter in its Fourier plane, the output being, after an optical inverse Fourier transform, the reconstructed distribution k(M). To synthesize the object filter it is necessary, the projections corresponding to Eqs.(1) or (4) being measured, to perform an analog back-projection producing a two-dimensional representation of each set of projections and then to realize physically, for each point M(x,y) of the tomographic plane, the summation of all the "straight lines" which represent the back-projections of all projections of the point M, θ varying from 0 to π .

4. Optical encoding of information in real time

Following the mathematical operations of back-projection and of summation, performed with a computer in transaxial tomography, we record, on a photographic plate, for each angle θ of projection, an amplitude grid corresponding to the scanning, the amplitude transmitted by each "line" of the grid being proportional to the integral(1) or (4). The grids are recorded in real time by means of an optical tracing system, a "light pencil" [2] , and we obtain directly, on the photographic plate, the object filter h(M), by superposition of the grids, step by step for θ varying from 0 to π. The "light pencil" is also used to produce the "ρ" filter. The principle of this technique seems to be very simple but, in practice, if we have to add a great number of grids, for example 36 grids corresponding to an increment of 5° for the angle θ of scanning, the recording on a usual photosensitive emulsion may be rather problematic. So we have preferred a sequential processing, each grid being filtered separately and the different outputs being added in proportion as the image of the tomographic plane appears on a photoemulsion or on a T.V. screen.

5. Sequential coherent optical processing with a ROACH filter

The principle of the "one-dimensional" sequential processing is based on the linear properties of the equations governing transaxial tomography, and that allows the order of operations to be reversed. Another advantage consists in using the same "ρ" filter as in the two-dimensional processing. The grids are recorded on a film (Fig.2), each one corresponding to a particular angle θ_i of scanning.

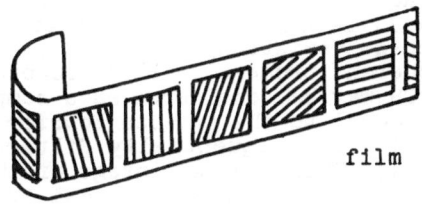

Fig.2. Recording of the grids for a sequential processing

The advantage of a sequential recording of grids is that the Fourier spectrum of each grid is one-dimensional and may be filtered by a "diameter" of the "ρ" filter, so that all the grids are filtered step by step with the same 2-D "ρ" filter (Fig.3)

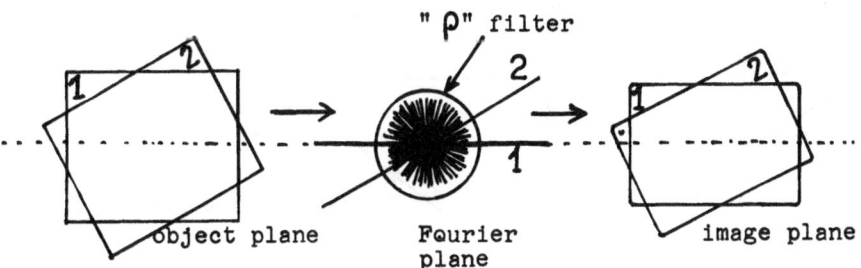

Fig.3. Sequential filtering of the grids with a "ρ" filter

The superposition of the filtered grids gives in real time, in a suitable viewing device, the 2-D reconstructed image. A fundamental drawback of this method is the well known difficulty due to phase variations which occur in positioning the filter. Therefore we have replaced the classical coherent set up by a more compact device using a R.O.A.C.H. filter.

The "Referenceless On Axis Complex Hologram" (ROACH) is a phase and amplitude filter which is an extension of the kinoform [4] . Compared to the kinoform, which gives on-axis phase information only, an holographic filter may be said "off-axis" because of the reference wave. To store amplitude information, CHU, FIENUP and GOODMAN [3] use multiple emulsions films, such as Kodachrom, in which different layers are exposed separately by lights of different wavelenghts. If, for example, we use the ROACH in red light, the amplitude variation is recorded by exposing the film with red light and the phase pattern is exposed with blue-green light. The red sensitive layer absorbs red while the other layers are transparent but cause variations in film thickness and refractive index.

Now, if we consider Eqs.(7), k(M) may be obtained by the equivalent operation of convolution:

$$k(M) = h(M) * T.F.^{-1}(\rho) \tag{8}$$

The convolution described by Eq.(8), in coherent light, may be simply performed by using a ROACH. The amplitude pattern corresponding to the "ρ" filter will be exposed with red light, for example, and the quadratic phase variation, modulo-2π , given by the lens of the classical coherent process, will be recorded with blue-green light. Such a ROACH may be considered as a complex lens with a point spread function equal to the inverse Fourier transform of ρ . The different grids are sequentially filtered in this compact coherent system (Fig.4). The great advantage of this method using a ROACH filter, compared to the classical Fourier filtering process, is to avoid problems of centering and positioning.

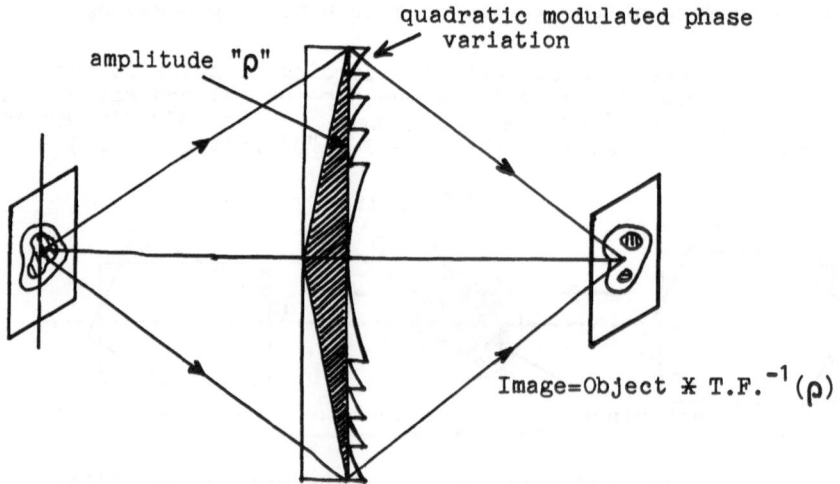

Fig.4. Coherent reconstruction with a ROACH

6. Conclusion

We have presented a simple analog method of transverse tomography which may be performed in real time without data storage and without long and expansive numerical treatments. This method may be applied to all sorts of measured quantities if they are linearly combined. The information(logarithms or square roots of detected energies, acoustic velocities,...) is encoded and preprocessed optically and the grids, recorded in real time, are sequentially processed in a simple optical device which may be completely automatized.

The operation of reconstruction described by Eq.(8), being a convolution, is practicable in incoherent light. With a one-dimensional correlator we produce the image of each grid on a mask which is translated in its plane at constant velocity. The product of both functions is integrated by a lens and, by means of an image memory, the result can be seen on a T.V.screen. The energy transmitted by the mask is the one-dimensional Fourier transform of the "ρ" filter. This function comprises positive and negative parts which are convoluted separately with the grid, the difference being achieved at the end.

This method may be extended to acousto-optics where the absorption coefficient is replaced by measured velocities. Applications to microwaves are also in study.

References

[1]   H.H. BARRETT and W. SWINDELL, Proc. IEEE 65 (1977) 89.

[2]   A. LANDRAUD, Optics Communications 25,n° 3 (1978) 305.

[3]   D.C. CHU, J.R. FIENUP and J.W. GOODMAN, Appl.Opt.12'(1973) 1386.

[4]   J.J. CLAIR, "Optical Holography" (1971) Plenum Press.

Optical Reconstructions from Projections via Deconvolution

J. Hofer

Lehrstuhl für Nachrichtentechnik, Technische Universität München
D-8000 München, Fed. Rep. of Germany

1. Introduction

The reconstruction of sectional images of an object from its projections is mainly important in medicine, e.g. in X-ray transaxial tomography. In the passed six years the reconstructions achieved there by digital computers have proved to give usefull new information. Since that time a lot of analog reconstruction methods have been proposed to use possible advantages like less cost or higher spatial resolution. A survey is given in the paper of PLATZER and GLÜNDER in these proceedings.

Almost all analog methods, which are comparable with the digital ones (i.e. the reconstruction of a sectional image is performed from projections only of the according object sectional slice) are essentially sequential methods. The sectional image is e.g. built up point by point. Thus, in optical reconstruction techniques mechanical movements are needed which are time consuming and demanding high accuracy. The only method that offers an optical parallel reconstruction was proposed by HANSEN and GOODMAN [1]. The reconstruction theory there leads on a laborious mathematical path, via circular harmonic expansion, to a spatial frequency filtering process combined with spatial weighting and coordinate transformations. Here, this is also achieved simply using methods of system theory introduced into optics long ago. It makes the filter evaluation more easily understood and offers the possibility to realize the filter e.g. by well known holographical techniques.

2. Theory

The theory of reconstruction from projections via deconvolution will be given here only very roughly. Detailed derivation and description can be found in [2].

If the sequential image is regarded as a two-dimensional density function and if its line integrals (projections) are arranged in the sinogram [3], this twodimensional arrangement can be derived from the sectional image by superposition of space-variant point responses. Just by a suitable coordinate transformation the superposition changes to convolution, i.e. each density function point gives then the same point response, only shifted according to the point position. So reconstruction means to deconvolve the distorted sinogram. This can be done, e.g. in parallel by coherent-optical filtering, e.g. using a holographically realized approximate inverse filter, since the point response is known and it is just a simple curve.

The result of deconvolution contains an approximation of the desired sectional image, which is still in the distorted coordinates and needs spatial weighting. Whereas the latter is optically simply to realize, the parallel inverse coordinate transformation has yet no feasible general solution for images with a high space-bandwith product. New optical means may be useful there, based on holographic pseudo-lens arrays, as described in [4], which were applied there to multichannel Fouriertransformations. Analog electronic means, working sequentially but fast, are also suitable for such coordinate transformations and may be already combined with the display. The result of deconvolution is only an approximation, because of errors due to truncation problems with the distorted functions and the approximate inverse filtering. Up to now it seems, that these errors can be kept in the range of those inherent in optical sinogram storage and optical computing.

3. References

1 Hansen,E.W.; Goodman,J.W.: "Optical Reconstruction from Projections via Circular Harmonic Expansion" Opt. Com., vol.24, no.3, March 1978 pp.268-272
2 Hofer,J.: "Optical Reconstruction from Projections via Deconvolution" Opt. Com. (in print)
3 Barrett,H.H.; Swindell,W.: "Analog Reconstruction Methods for Trans-axial Tomography" Proc.IEEE, vol.65, no.1, January 1977, p.101
4 Platzer,H.; Glünder,H.: "Generation and Use of Local Power Spectra by Coherent Optics" Inst. of Physics Conference Series, no.44, "machine-aided image analysis 1978"

VI. Holography in Ophthalmology

Application of Holography in Ophthalmology

H. Ohzu[1] and T. Kawara[2]

[1]Department of Applied Physics, Waseda University
Ohkubo 3-4-1, Shinjuku, Tokyo 160, Japan
[2]Department of Ophthalmology, Tokai University
Bohseidai, Isehara, Kanagawa 259-11, Japan

1. Introduction

Holography is a unique optical technique which has a number of advantages
over the conventional photography. With holography, both the amplitude and
phase are stored in a photographic emulsion called a hologram. The practical
processes and mathematical principles involved in holography have been presented elsewhere [1]. Since the hologram records all the information contained
in a wavefront from the object, the reconstructed images by means of holography are truly three-dimensional and extremely realistic. It is reasonable
to apply the holographic technique to investigation of the eye. Holograms
can be very useful in Ophthalmology because of their enormous capacity for
information storage and reconstruction.

2. Advantages of Holographic Techniques in Ophthalmology

A hologram can provide a three-dimensional image of the entire eye or of
several parts of the eye. One can focus separately and successively on the
different planes of depth throughout the eye from the examination of the reconstructed image [2]. If the reconstructed image is examined with an optical
system such as a microscope of sufficiently high magnification that has
only a limited viewing field of depth, one can focus separately on different
layers of the retina or other tissue, studying it plane by plane, just as a
histopathologist uses serial sections to examine a preserved eye [3]. The
site of abnormalities such as retinal detachment or intraocular foreign bodies
is also very easily detected and determined. It is possible to measure pathological abnormalities such as cupping of the optic nerve head, melanomas or
neovascularization within the vitreous [4].

A hologram made with a single exposure of the eye can provide a permanent
record of the structure of the entire eye. With a single hologram, any desired
layer can be examined at any later time on subsequent reconstruction. Holography, therefore, allows detailed retrospective study of portions of the eye
not examined initially. This is very advantageous especially in clinical ophthalmology because of the following reason. In ordinary photography, one must
predetermine which portion of the eye is to be photographed, i.e., the cornea,
the iris, the anterior or posterior surface of the lens, or the superficial
layers of the retina. The portion is, then, focused and photographed using an
instrument (such as fundus camera or photo-slit lamp microscope) to record it.
At a subsequent time, one may need a photograph of a different part of the
same eye; this may not be available [4].

Other holographic applications have been also proposed. It will be possible by means of holography to get more precise information than by using ordinary ophthalmic instruments about the optical constants of the eye and configuration changes accompanying accommodation or changes in intraocular pressure [5]. With a special holographic technique, it is theoretically possible to obtain an extremely high resolution of about 1 μm in the reconstructed retinal image [5]. Using this method, the smallest cones of about 1.5 μm in diameter might be visualized.

Using a technique of two-frequency holographic interferometry, one can form fringes of periodic intervals corresponding to the depth on an object's surface. The contour map can then be generated on various parts of the eye, notably the retina, and small differences in height of its surface would be detected. The cat's retina has been contoured using this method with resolution of 0.46 mm in depth [6].

The double-exposure interferometry of holography allows displacements between an original state and a changed state to be measured. One can precisely determine the deformation of various parts within the eye by this technique, though it is not applied so easily to a living eye-ball for the reason of difficult experimental conditions. The scleral deformation of the extracted eye accompanying changes in intraocular pressure has been measured using this technique [7].

A computer-generated hologram can produce an arbitrary wavefront. A new method has been proposed for measuring refraction of the eye using such a hologram [8].

3. Holographic Recordings of the Eye

VAN LIGTEN et al first suggested the basic application to the ophthalmic field [9]. They published holograms of model eyes only, using a lens to form an image hologram. They stated that the resolution in the reconstructed image of the fundus was relatively poor which was due to the deteriorated state of the wet model eye employed [9].

The earliest published record of holograms of an eye of a living animal (cat) is that by CALKINS and LEONARD [2]. As a more practicable way, WIGGINS et al improved a conventional fundus camera and constructed a hologram of a desired area in a cat's retina [3,10]. A third technique makes use of a combination of optical fiber coupled to a contact lens. The holographic images of a living cat's retina produced with this method were reported by us [11,12,13], and that of a living rabbit's retina was produced by ROSEN [14]. Recently, TOKUDA et al reported on a holocamera for three-dimensional microscopy of the unanesthetized human eye [15].

In the following sections, the several optical systems developed by many authors for constructing the ocular hologram and the results obtained with each method are discussed in some details.

3.1 Holography Applied to the Living Eye

CALKINS and LEONARD [2] used a very orthodox method in holography. Their experiment was arranged on an optical bench and they used a continuous wave He-Ne laser with an output of 58 mW at 632.8.nm wavelength. Their animal selected was a cat, which was anesthetized and its head was held firmly by an aluminum jig to stabilize the motion during the exposure. The hologram of the cat's eye was recorded on an Agfa Gevaert 10E70 emulsion plate which had been hypersensitized in 2% triethanolamine. After an exposure of 1/250 second, it was developed 12 minutes in Agfa Metinol U and bleached.

They stated that the energy density at the retina was 1,200 ergs/cm^2. Its density level was more than 5×10^3 times below the standard given by the Department of Health, Education and Welfare (HEW) guidelines for threshold for detectable retinal damage in humans with the pulsed non Q-switched laser irradiation in the millisecond range. Recommended safe levels were chosen two orders of magnitude lower than the damage threshold level, which was still more than 50 times greater than their light energy [2].

The interesting technique of their method is the use of a corneal plano-concave contact lens and two orthogonal linear polarizations for the reference and illumination beams. The undesired specular reflections from the eye and from the contact lens surface retained the same linear polarization as the light incident on the eye but were at right angles to that of the reference beam. Since only the light waves polarized in the same direction can create interference fringes, the specular reflections will not contribute to forming the hologram. The diffuse reflection produced by the retina or the iris emerges as depolarized, and an interference pattern could be satisfactorily created on the emulsion plate. The contact lens placed on the cornea was also useful to prevent corneal drying and to neutralize the refractive power of the cornea so that a larger area of the retina was recorded.

Reconstructed real images of the iris and of the retina were formed by illuminating the developed hologram with a converging wavefront of laser light. These two images which planes were mutually much different in depth were reconstructed from a single hologram; they actually showed the characteristic three-dimensional nature of holography. The optic disc and some fine vessels were clearly recorded, but no measurement of their diameter was given. Depth within the retina was not demonstrated.

3.2 Holography Using a Fundus Camera

In 1972, WIGGINS et al reported on holograms of the retina of an anethetized cat using a continuous wave argon ion laser with a prism wavelength selector for operation at the green wavelength of 514.5 nm [3,10]. An etalon was also used to give coherent length extension of over ten meters. They constructed the hologram using a modified Zeiss fundus camera. The electronic flash was replaced by a laser beam for illumination of the fundus and the beam was focused in front of the anterior corneal surface. The camera back was removed and the holographic emulsion placed just in front of the plane of the fundus image formed by the camera optics. Detailed modification of the Zeiss fundus camera are described elsewhere [10].

Their exposure times were 6 to 8 msec with using of unsensitized emulsion plate of Agfa Gevaert 10E56. Development was in Agfa Metinol U for five minutes and the plates were not bleached. The retinal energy densities in their experiment were calculated to be 2×10^4 ergs/cm^2 to 9×10^4 ergs/cm^2. Although these values are much higher than those of CALKINS and LEONARD, they are nevertheless below the HEW guidelines which place the threshold for retinal damage as 5×10^6 to 1×10^7 ergs/cm^2.

The fundus images were reconstructed by illuminating the hologram plate with a laser beam identical to the construction reference beam. The real images were magnified by a conventional microscope, while the recorded area was smaller than that obtained by CALKINS and LEONARD. For some photographs of holographic images, they placed a rotating diffuser in the plane of the image upon which the examining microscope was focused. This method reduces the grainy speckle noise which is due to the random interference of the laser light. Two photographs were proposed to demonstrate the three-dimensions within-in the retinal layers themselves. These were taken through the examining microscope which was focused on the superior vessels and on the inferior vessels respectively. They stated that these layers were separated in depth by 120 μm, and that the smallest vessels recorded in the holographic image were about 30 μm in diameter [3]. It was calculated by comparing the holographic image with a standard fundus photography and a fluorescein angiography obtained from the same fundus. The value of 30 μm was thought to reach very nearly the resolution limit of the fundus camera.

The advantages of their method are the ability to determine the exact area of the fundus of which the hologram is to be made, and easy adaptability of existing ophthalmic equipment to make holograms. However, the recorded area of the fundus with an single exposure and the resolution of retinal vessels are limited by the optical performance of the modified fundus camera. It is not easy to record the information about the other ocular parts (such as the iris and the lens) being separated largely in depth from the fundus plane.

In the holographic fundus camera produced by WIGGINS et al, the energy density probably becomes extremely high where the laser beam for illumination is focused and concentrated with an approximate Maxwellian view. It was reported by NASH et al that the corneal endothelial damage was caused following the focal laser intervention [16]. The cornea, therefore, may be in peril of damage by this high energy density. If the subject's eye moved just before the exposure, it would damage the iris or the sclera. It is advisable to avoid any concentration of the laser beam even at the cornea.

3.3 Ocular Holography Using Fiber Optic Illumination

On ocular holograms produced by a third optical system were reported by us [11,12,13] and by ROSEN [14]. The fundus was illuminated through a combination of optical fiber bundle coupled to a contact lens. This method permits simple wide angle observation of the fundus. The equator plus fundus camera has been already developed by POMERANTZEFF [17]. It is very advantageous to record the wide angle fundus image with holography, because the curved image plane of fundus cannot be easily focused onto a flat-film plane with conventional photographic techniques.

In this section, we describe the detailed exposition for our experiment [13], in which the images of cat's fundus and its iris were recorded on the hologram by using fiber optic illumination and the retinal images separated in depth by about 75 μm were reconstructed respectively on a thin film of liquid crystal forming a screen, with the resolution of about 10 μm.

3.3.1 Experimental Procedure

We sought to take a fundus hologram of a living eye using the simplest configuration. The major problem encountered in our holographic experiments of the outer or inner eye in the living animal was that the object to be recorded could not be moving. This means one had to use either very short exposures or stabilize optically the system vis-a-vis the recording plate, in which the latter method was used now. A special system has been developed for producing a hologram of a living eye. The laser light for illumination was introduced into the eye by means of a flexible optical fiber bundle as schematically shown in Fig.1. The end of this bundle was aligned annularly and attached to a planoconcave contact lens. The fundus was viewed through its central part with an aperture of 6 mm. Each fiber had a diameter of 30 μm. The other end of the fiber bundle, the input terminal for the illumination light, is bundled circularly with a diameter of about 6 mm. Fig.2 shows the apparatus which was made to be very compact. A mechanical support held the apparatus of the optical fiber system coupled to the contact lens stationary; the lens in turn helped to maintain the position of the subject's eye and to stabilize it during the recording.

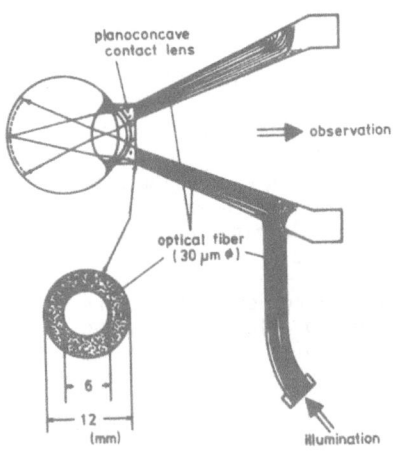

Fig.1 Schematic diagram of the optical fiber coupled holographic apparatus. The eye's structures are illuminated by a ring of fibers and viewed through the central part. From KAWARA and OHZU [13]

Fig.2 Picture of the optical fiber coupled holographic apparatus. From KAWARA and OHZU [13]

Holograms were taken of a cat's eye, because its fundus with tapetum reflects a good deal of light. The cat was generally anesthetized with ketamine hydrochloride. A retrobulbar block was performed with 2% lidocain hydrochloride to stop eye movements during exposure to the light. Two drops of tropicamide were instilled in the eye to dilate its pupil. Hydroxy-ethylcellulose was applied to the contact lens before attaching it to the cornea to protect its epithelium.

Figure 3 represents a schematic diagram of the optical system for producing the hologram. A continuous wave argon ion laser was used with an etalon which extended its coherence length. The wavelength of 514.5 nm was selected to enhance the contrast of the retinal image. The laser beam with an output of 100 mW was split into two beams by a beam splitter. One beam was directed onto the input terminal of the optical fiber bundle in order to illuminate the ocular structures diffusively. An object wave which was the light scattered from the eye's structures reached the hologram plate through a non-spherical projection lens. This lens formed the fundus image, which was monitored prior

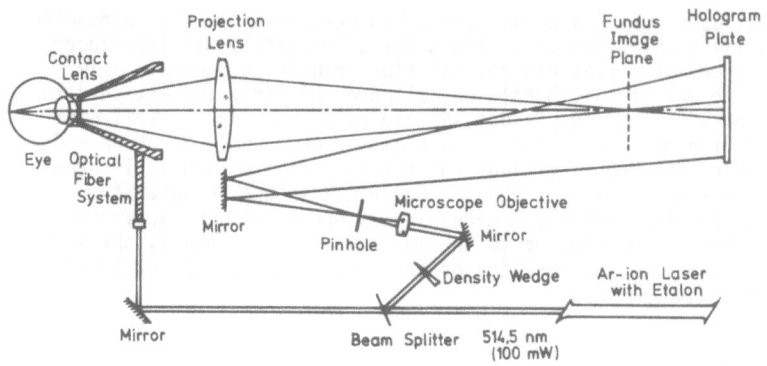

Fig.3 Ocular hologram recording geometry by use of the optical fiber coupled apparatus. From KAWARA and OHZU [13]

to exposure, on the plane shown as an interrupted line in Fig.3. The other beam reflected from the beam splitter was used as a reference wave in the form of a spherical wavefront, which was incident on the hologram plate via a small surface mirror mounted beside the projection lens.

Glass plates with an Agfa Gevaert 10E56 emulsion were used without pre-sensitization. The intensity ratio of the reference to object wave at the hologram plate was varied between 2:1 and 5:1 by use of a neutral density wedge. Exposure times were 8 to 16 msec. The energy densities at the retina were calculated to be between 1,500 ergs/cm² and 3,000 ergs/cm². The plates were developed in Kodak D-19 for 5 to 7 minutes according to the conventional schedule in photography and bleaching was not performed.

To reconstruct the ocular image from the developed hologram, a diverging wavefront of laser light was made incident on the emulsion side of the holo-gram. A reconstructed virtual image was observed or photographed through the hologram. Another method was introduced to reduce a grainy speckle noise which is caused by the extremely good coherence properties of the laser light.

Figure 4 shows the schematic diagram of the optical system used for this purpose. The hologram was illuminated with a converging wavefront of laser light. In this case, a real image was reconstructed directly onto the plane where a thin film screen of liquid crystal was set up. A nematic liquid crys-tal of compound MBBA (methoxybenzylidene-p-n-butylaniline) was sandwiched

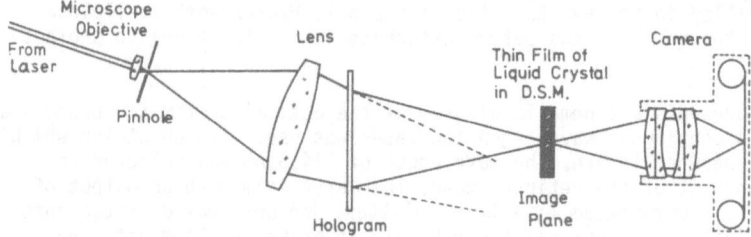

Fig.4 Optical arrangement to reconstruct the holographic retinal images on the liquid crystal screen in DSM. From KAWARA and OHZU [13]

between nesa-coated electrodes using Teflon spacers of 30 μm thickness. In its quiescent state with no electric field applied, the thin film screen of the liquid crystal was essentially transparent. When a static electric field of about 50 volts was applied between those electrodes, the liquid crystal became turbulent and scattered the light. In this state, which is designated as the "Dynamic Scattering Mode (DSM)" [18], the liquid crystal appeared white and could be used for the light screen. The Brownian motion of its scattering center diffused the laser light incoherently and reduced the speckle noise appearing at reconstruction. The image reconstructed on the liquid crystal screen was viewed from behind or photographed by use of a conventional macro-camera. To examine the retinal image throughout its depth with a given mag-nification, the camera and screen were moved along the optical axis tandemly so as to maintain constant separation between them.

3.3.2 Results and Discussion

Figures 5 and 6 are typical photographs of virtual images reconstructed from a developed hologram. In Fig.5, a fundus image of rather high quality is clearly visible along with the optic nerve disc and fine retinal vessels. The field of view is wider than that obtained by other methods described above. The recorded retinal diameter seen in Fig.5 is about 6 mm which is about twice that recorded by WIGGINS et al [3,10]. There is, however, much grainy speckle noise which is due to random interference of laser light. This speckle pattern lowers the resolution and may become a problem in the case of high magnification. Fig.6 shows a part of the mydriatic iris, which was

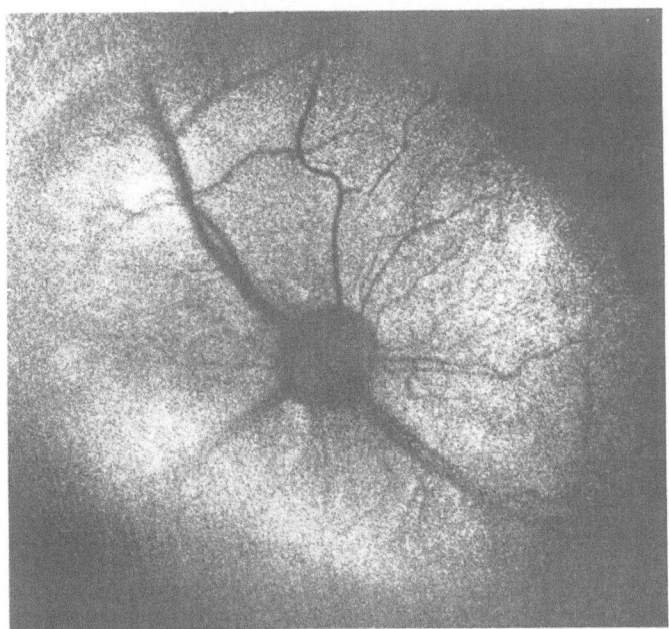

Fig.5 Reconstructed holographic image of a cat's retina using a fiber optic illumination. The virtual image was reconstructed by a diverging wavefront. From KAWARA and OHZU [13]

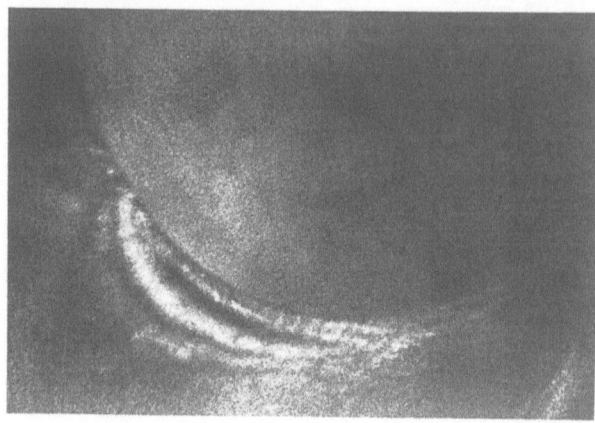

Fig.6 Reconstructed holographic image of the iris. This was photographed from the same hologram as Fig.5 by focusing on a different surface. From KAWARA and OHZU [13]

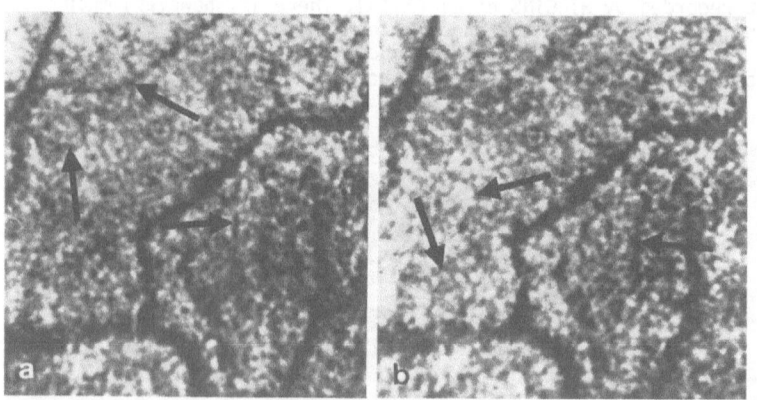

←500 μm→

Fig.7 Magnified real images reconstructed on the liquid crystal screen, taken in the way described in Fig.4. (a) Superficial vessels in focus. (b) Deeper vessels in focus. These vessels were separated in depth by 75 μm. The typical ones are indicated by arrows. From KAWARA and OHZU [13]

photographed simply by moving the camera lens so as to focus the plane closer to the hologram. These photographs have been obtained from a single hologram in order to demonstrate the characteristic advantage of the three-dimensional recording and the enormous field of depth which is contained in the hologram.

Figure 7(a,b) shows magnified real images of the peripheral retina reconstructed from a hologram on the liquid crystal screen to reduce the speckle noise. These were photographed in the way described in Fig.4 in which two retinal layers were separated throughout in depth. Fig.7(a) was obtained by placing the liquid crystal screen in the plane where the vessels in the superficial layer were in focus, while Fig.7(b) was taken where the vessels in the deeper layer were in focus. The larger vessels are clearly visible as bold lines in both figures. One can see the smaller vessels here and there. The finest vessels resolved at different quarters are thought to be capillary vessels with a diameter which was calculated to be 10 μm at the most. The magnification is represented as a scale at the lower part of the figure. Furthermore, there are some superficial vessels recognized only in (a); the typical ones are indicated by arrows. Those vessels are not found in (b). On the other hand, the arrows in (b) indicate the deeper vessels which cannot be distinguished in (a). The distance in depht can be estimated from the relationship between lateral and longitudinal magnification in holography. It was, therefore, concluded that the superficial and deeper layers within the retina shown in Fig.7(a,b) are separated in depth by about 75 μm.

The depth interval between layers in the retina has been measured with the screen of liquid crystal in DSM. The liquid crystal screen in this state made the light incoherent and as a result reduced the speckle noise. A rotating ground glass plate also diffuses the light incoherently, easily and effectively [3,10]. This method, however, must be carried out by use of mechanical motion. We have been concerned with the oscillation of such a diffuser's surface, which accompanies mechanical rotation. The oscillation might result in a reduction of the resolution in depth by extending the depth of image focused on this diffuser. On the contrary, the liquid crystal screen of 30 μm thickness is perfectly stable. For the purpose of estimating the accurate distance in depth, it might be adequate to use the liquid crystal screen instead of the rotating diffuser.

In the two ways described above, only one kind of speckle noise is averaged out by the movement of the scattering center within the liquid crystal or diffuser. Such speckle is due to random interference of laser light from the optical elements used at reconstruction. The other is caused in taking the hologram itself. This speckle is due to random interference of the light reflected from the fundus. Usually the larger the aperture of the optical system, the smaller is the size of speckles. In our experiment with a 6 mm aperture, the blood capillary has been resolved to be 10 μm in diameter as shown in Fig.7. This value is believed to be a resolution limit determined from a line spread function or modulation transfer function of the eye. It is also the satisfactory value in comparison with an ordinary fundus photograph. CALKINS [6] estimated a resolution limit of 19-20 μm from the line spread function measured in normal eyes at 6.6 mm dilatation by CAMPBELL and GUBISCH [19]. He assumed that the resolution limit was given when one selected 50% as the desired contrast transmittance, because the retinal detail had rather low contrast. In our experiment, better resolution has been achieved without enhancing the retinal contrast such as fluorescein angiography. It is probably because the contrast of the cat's fundus illuminated with 514.5 nm wavelength was higher than that assumed by CALKINS and that the line spread function of the cat's refractive system including the contact lens was narrower in bandwidth than that measured by CAMPBELL and GUBISCH.

OHZU proposed an optical principle for holographic observation of individual receptor cells in the living retina [5]. It may be indispensable to eliminate the speckle noise in order to observe these cells, i.e. cones or rods. To further minimize the speckle noise, the flexible fiber bundle for illumination would be vibrated at one end, the input terminal, during the exposure. Such an average as to keep the laser light coherent would give rise to satisfactory reduction of the speckle noise.

The most significant problem involved in applying this study to the human eye is the energy level. In the transparent ocular structures, laser light may produce damage where the light is focused in the eye. The illumination by means of the optical fiber bundle makes no concentration and has an advantage in this regard . The light was introduced through a ring of fibers in our system, and consequently the energy density was very low even at the cornea. Therefore, it is not necessary to discuss the energy density in the transparent media. As for the retina where the damage is liable to be sustained, Department of Health, Education and Welfare (HEW) guidelines place safe exposures between 50,000 ergs/cm^2 and 100,000 ergs/cm^2. In our experiment, the maximum energy density on the retina was calculated to be about 3,000 ergs/cm^2 which is well within the limit of safety. No unusual retinoscopic changes on the cat's retina could be found after many exposures. However, we did not examine its retina with an electron microscope. Using an electron microscope, ADAMS et al found a disarray of photoreceptors in several monkeys exposed with extremely low levels [20]. Since the human retina reflects light less than the cat's retina, a higher energy of illumination would be required.

3.3.3 Fundus Holography Using Special Wide-Angle Contact Lens

ROSEN reported on fundus holograms of the anesthetized albino rabbit [14]. His method for constructing the hologram was similar to ours, but he used a special wide-angle contact lens coupled to fiber optic assembly designed and developed by POMERANTZEFF [21].

A continuous wave argon laser with an output of 210 mW at 514.5 nm wavelength was used as the light source. The retinal energy density was approximately 0.15 mJ/cm^2 with an 8 msec exposure time. Agfa Gevaert 10E56 film plates were used with a five minute development time in Kodak D-19. One noted that the field of view obtained was somewhat in excess of twice the 30° field obtained with the Zeiss fundus camera.

3.4 Holography Applied to the Human Eye

Recently, a holographic application to the living human eye was reported by TOKUDA et al [15]. They developed a holocamera which safely recorded holograms of human and rabbit eyes. Except for dilating solution and corneal anesthetic, the human eye was unanesthetized. Anesthetized rabbits were also used to follow the development of microwave induced cataracts.

An anti-reflection coated planoconcave corneal contact lens was used to negate the focusing effects of the cornea. The effect of specular reflections was reduced by orthogonally polarizing the reference and the eye illuminating beams. A fail-safe interlocking shutter system kept exposures within the ANSI safety standards. This system synchronized a single window chopper with a double-bladed electromechanical shutter. The three-dimensional information of the entire useful depth of the eye could be stored with an exposure time of 0.3 msec.

The reconstructed real images were projected on a vidicon faceplate of a closed-circuit television system, enabling convenient scanning in X-Y-Z-dimensions of the reconstructed eyeball. Several comparisons were made with a slit lamp microscope and fundus camera. As a result, TOKUDA et al stated that the ability to scan in three dimensions would help researchers follow intraocular irregularities, such as cataracts or glaucoma, in addition to written records, sketches, or two-dimensional photographs [15].

4. Other Applications of Holography in Ophthalmology

4.1 Two-Frequency Holographic Contour Generation on the Retina

A contour map is generated by two steps in holography. An object firstly is recorded on a hologram with two discrete wavelengths (λ_1 and λ_2) simultaneously. Then the hologram developed is reconstructed by only one of their wavelengths. The wavelength used for reconstruction produces a magnification (or reduction) in the portion of the image recorded with the other wavelength. The separate wavefronts, corresponding to each image, interfere forming fringes at periodic intervals of depth. The distance between two adjacent fringes corresponds to $(\lambda_1\lambda_2)/2(\lambda_1-\lambda_2)$.

Holographic contouring of a cat's fundus was reported by CALKINS [6]. The wavelengths selected were the 476.5 nm line from an argon laser and the 476.2 nm line from a krypton laser. The contour interval obtained was 0.46 mm. Agfa Gevaert 8E56 emulsion was used to construct a volume hologram. Although this method produces satisfactory results, the contour interval of 0.46 mm is thought to be too large to measure the small unevenness such as the cupping of the optic nerve head.

4.2 Measurement of Deformation of the Eye Accompanying Changes in Intraocular Pressure Using Holographic Interferometry

Holographic interferometry permits measurements of minute changes in shape of three-dimensional complicated shapes. This technique, therefore, may well provide an exceedingly useful tool in examining changes within the eye-ball such as shrinkage of the vitrious and development of cataract, of vitreal strands, and of retinal edema [4]. Practically, it is not easy to apply this technique in measuring deformations of a living eye which may move. For holographic interferometry, the eye to be examined must not move between two exposures in a double-exposure technique. As to real-time holography, its application may be even more difficult because during reconstruction both the eye and the processed hologram must be replaced in exactly the same position they occupied during constructing the hologram. It is believed that such optical condition can hardly be satisfied, as far as a living eye is used.

MATSUMOTO et al measured the deformation of the eye accompanying changes in intraocular pressure (IOP) , by using extracted eyes of white albino rabbits [7]. The deformation of the sclera at the rear of the eye was measured for various small incremental values of IOP using double-exposure holographic interferometry. The visual function and size of the eye are seriously affected by changes in IOP. So, quantitative measurements of the deformation of the eye caused by changes in IOP are very important in ophthalmology, especially in investigations of pathological conditions such as glaucoma, ametropia, and those following intraocular surgery [7].

A special chamber was developed by them in order to hold the extracted eye and to observe the deformation of the sclera. After a Teflon tube was inserted into the anterior chamber of the eye at the limbus, the eye was cemented to the tapered hole of a metal plate and then the Teflon tube was tightly attached to the guide hole in it. A physiological salt solution filled the chamber fitted with a flat glass observation window. The Teflon tube was connected by a long vinyl tube to a reservoir containing the physiological salt solution. This reservoir could be raised and lowered to act as a manometer, which changed the IOP to any desired value. The deformation of the sclera at the rear of the eye was measured through the observation window by using the double-exposure technique. The holographic scleral image was reconstructed with interference fringes superimposed on it according to its deformation. A relative displacement of the sclera was calculated from the fringe pattern. Each generated fringe corresponded to a displacement of 0.25 μm along the visual axis (front to back) of the eye.

The method using holographic interferometry is very useful to measure the deformation of the eye. Though the technique used bei MATSUMOTO et al. was only applied to the extracted eye, this method is believed to be very advantageous in investigating the glaucoma experimentally, especially in estimating the relationship between the intraocular pressure and the cupping of the optic nerve head which probably correlates to the deformation of the sclera at the rear.

4.3 Refratometry by Using a Computer Generated Hologram

It is very important to measure accurately the refractive power of the eye for investigating the accommodation and the anomaly of refraction such as hymyopia, and astigmatism. It is also indispensable to determine the lens power of glasses or contact lenses best suited for correction of these abnormal eyes. So, many refractometers have been developed up to now and are utilized effectively in clinical ophthalmology and in optometry.

A new method was proposed by us for measuring the refraction subjectively [8]. The interference fringe pattern of equally spaced concentric circles, which is reconstructed from a computer-generated hologram, is used as a viewing target for measurement. This test pattern is projected on to the retina so as to be modified differently according to various refractive states of the eye. For example, the diameter of every circle changes in case of hyperopia and myopia, and ellipses are observed by an astigmatic eye. A similar fringe pattern as mentioned above is additionally projected with a Badal lens which is placed at its focal distance in front of the eye. This additional reference pattern, which is also reconstructed from the hologram, is focused at the nodal point of the eye and diverges to project the reference circles on the retina without being influenced by the refractive state. The diameter of this reference circles can be controlled by converting the power of the Badal lens. As a result, the test and the reference circles are superimposed on the retina and hence Moiré fringes are generated, which show various patterns according to the refractive state and the power of the Badal lens. Therefore, one can decide the refraction subjectively by means of interpreting the Moiré fringe pattern.

The equally spaced concentric circles were produced by the interference of two conical waves being collinear in direction and different in slope to each other. These conical waves were reconstructed simultaneously from the computer-generated hologram. A binary synthetic hologram was made using the method developed by Lee [22]. This hologram having 100 fringes was plotted on a X-Y graphic plotter and photoreduced on a emulsion plate to a practical size of 3.7 mm square. Fig.8 shows the schematic diagram of the optical system for measuring the refraction of the eye. The computer-generated hologram was illuminated by a collimated laser beam and reconstructed interference fringe circles. A spatial filter was placed in the focal plane of the lens to block off the undesired diffracted waves. The test circles were projected through the lens fixed in place and in power. The reference circles were projected through a zoom lens used as a Badal lens.

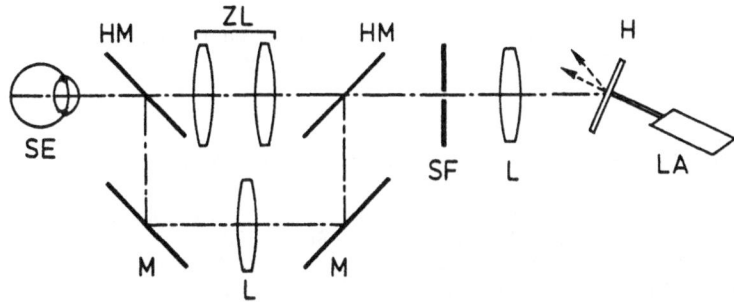

Fig.8 Optical arrangement for measuring refraction of the eye. HM,half mirror; L,lens; M,surface mirror; SF,spatial filter; H,computer-generated hologram; LA,He-Ne laser; ZL,zoom lens as a Badal lens; SE,subject's eye. The subject converts the power of the zoom lens so that Moiré fringes are observed. From SAITO et al [8]

Figures 9 and 10 show the results obtained from preliminary experiments, in which the cylindrical lens was preset in front of the subject's eye with normal refraction. Fig.9 shows the correlation between the cylindrical axis of the preset lens and that measured by the new method. The measured values almost coincide with the preset values as expected. The measurement errors of the axis against the cylindrical power are presented in Fig.10.

Interference fringes are generated in all areas where the two wavefronts are in existence simultaneously. Thus the interference pattern projected on the retina is hardly blurred in any refractive states, also the Moiré fringe pattern. Such test pattern (Moiré fringes) used in this method is very different to the viewing target used in ordinary subjective refractometry. Accordingly it might allow the effect of accommodation to be smaller than that caused in the method that minimizes the blur of the target illuminated by an incoherent light.

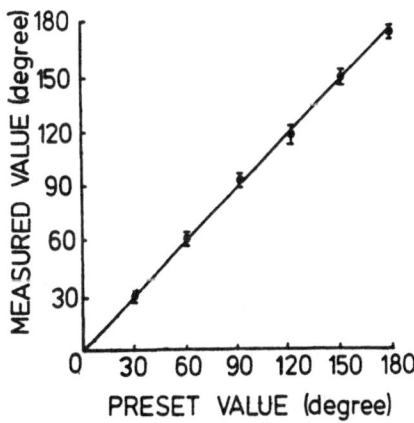

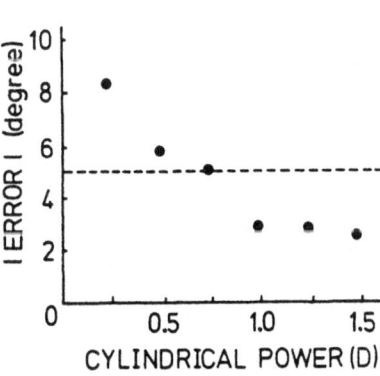

Fig.9 Correlation between the cylindrical axis of the preset lens and that measured by the new method. From SAITO et al [8]

Fig.10 Measurement errors of the cylindrical axis against its power. From SAITO et al [8]

We are now preparing reports on the other components of refraction such as the spherical power and the cylindrical power and on the application of this method to clinical ophthalmology and optometry.

5. References

1. G.Stroke: Introduction to Coherent Optics and Holography. Academic Press, 1966, New York
2. J.L.Calkins and C.D.Leonard: Holographic recording of a retina using a continuous wave laser. Invest.Ophthalmol., 9:458-462, 1970
3. R.L.Wiggins, K.D.Vaughan and G.B.Friedmann: Fundus camera holography of retinal microvasculature. Arch.Ophthalmol., 88:75-79, 1972
4. K.D.Vaughan and R.A.Laing: Holography of the eye: A critical review. in Laser Applications in Medicine and Biology, Edt. M.L.Wolbarsht, 2:77-132, 1973, Plenum Press, New York
5. H.Ohzu: Proposed applications of holographic techniques to the optics of the eye and vision research. in Application of Holography, Edt. E.S. Barraket et al, 365-376, 1971, Plenum Press, New York
6. J.L.Calkins: Fundus camera holography. in Holography in Medicine, Edt. P.Greguss, 85-89, 1975, IPS Science and Technology Press, Surrey
7. T.Matsumoto, R.Nagata, M.Saishin, T.Matsuda and S.Nakao: Measurement by holographic interferometry of the deformation of the eye accompanying changes in intraocular pressure. Appl.Optics, 17:3538-3539, 1978
8. K.Saito, T.Yatagai and H.Ohzu: Proposed application of a computer-generated hologram to refractometry. Preprint of the annual meeting of the Jpn.Soc.Ophthalmol.Optics, 1977, p.8
9. R.F.Van Ligten, B.Grolman and K.Lawton: The hologram and its ophthalmic potential. Am.J.Optom., 43:351-363, 1966
10. R.L.Wiggins, K.D.Vaughan and G.B.Friedmann: Holography using a fundus camera. Appl.Optics, 11:179-181, 1972
11. T.Kawara, H.Ohzu, Y.Sugimachi and A.Nakajima: Holographic recording of the living cat's retina. Preprint of the meeting of the Jpn.Soc.Appl. Phys., 1973 Autumn, p.65
12. H.Ohzu: Holographic Ophthalmometry. in Holography in Medicine, Edt. P. Greguss, 82-84, 1975, IPS Science and Technology Press, Surrey
13. T.Kawara and H.Ohzu: Fundus holography using fiber optic illumination. Jpn.J.Ophthalmol., 21:287-296, 1977
14. A.N.Rosen: Fundus holography through a wide-angle contact lens. Invest. Ophthalmol., 12:786-788, 1973
15. A.R.Tokuda, D.C.Auth and A.P.Bruckner: Development of a holocamera for 3-D microscopy of the unanesthetized human eye. J.Opt.Soc.Am., 68:1382, 1978
16. J.P.Nash, M.G.Wickham and P.S.Binder: Corneal damage following focal laser intervention. Exp.Eye Res., 26:641-650, 1978
17. O.Pomerantzeff: Theory and practice of the equator plus camera (EPC). Preceeding of Int.Sym.Ophthalmol.Optics, May 7-9 1978, Tokyo, pp.117-119
18. G.H.Heilmeier, L.A.Zanoni and L.A.Barton: Dynamic scattering: A new electrooptic effect in certain classes of nematic liquid crystals. Proc.IEEE, 56:1162-1171, 1968
19. F.W.Campbell and R.W.Gubisch: Optical quality of the human eye. J.Physiol., 186:558-578, 1966
20. D.O.Adams, E.S.Beatrice and R.B.Bedell: Retina: Ultrastructural alterations produced by extremely low levels of coherent radiation. Science, 177:58-60, 1972
21. F.Govignon and O.Pomerantzeff: Wide-angle holography of the eye. Trans. Am.Acad.Ophthalmol.Otolaryngol., 76:1214-1220, 1972
22. W-H.Lee: Binary synthetic holograms. Appl.Optics, 13:1677-1682, 1974

VII.   Holography in Urology

Interferometric Investigations of the Rabbit Urinary Bladder.
I. Holographic Registration of Bladder Deformations in vitro

K. Grünewald and H. Wachutka

Research Department Dornier System GmbH
D-7990 Friedrichshafen, Fed. Rep. of Germany

A. Hofstetter and R. Böwering

Department of Urology, Hospital
Thalkirchner Straße, D-8000 München, Fed. Rep. of Germany

1. Introduction

The blind biopsy applied frequently today as a method of planning bladder-
operations is unsatisfactory especially because it is accompanied by
additional physiological stress and therefore by an additional risk for
the patient. The aim of this examination, the early recognition of
cancerous indurations of the submucosa and the quantitative determination
of their extension in the bladder wall, can not be achieved by classic
endoscopy but it could probably be done by combining endoscopy with a
suitable method of strain-stress registration as e.g. holographic interferometry. Applying this it should be possible to deduce the extension of
areas and the degree of the elastic tissue alterations in these areas
from the interference pattern of the pressure loaded bladder.

The aim of our work is the development and the clinical test of an
endoholoscope. In the following it is reported on first in vitro
investigations which should show the feasibility of taking interference
patterns of a bladder with a certain reproducibility and of detecting
artificial tissue alterations on the fringe pattern when a suitable load
is applied.

2. Experimental Arrangement and Results

The experimental arrangement is shown in Fig.1.

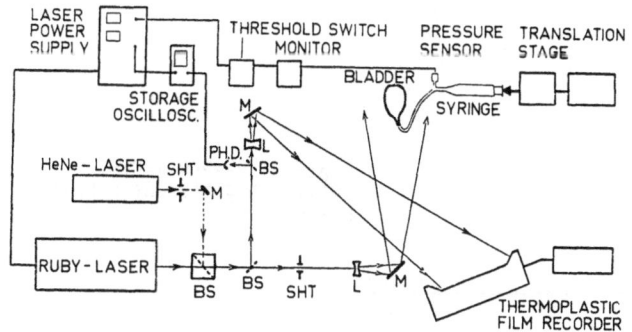

Fig.1 Double-impulse-holography of pressure loaded bladders

147

Holography is carried out using a double-impulse-ruby-laser K 1 QDH of korad the pulseduration of which is 25 ns with an interval between both pulses adjustable from 0.1 to 1.2 ms. The wavelength is 0.694 µm and the energy 50 mJ per Puls.
The laserbeam is splited by a beamsplitter BS into an illumination and a reference beam which are collimated by the lenses L and led to the object and to the hologram respectively via mirrors M. The illumination beam is reflected diffusely by the bladder onto the hologram which is registered on the thermoplastic film of the instant image device HSB 1000 (Rotten-kolber). The advantage of this registration is that the holograms may be reconstructed within 5 s after exposure without any chemical processing. For beam adjustment and hologram reconstruction a HeNe-laser is used the light of which is switched on and off by a shutter SHT. During recon-struction the illumination light pass is blocked by another shutter. For the control of pulse height and pulse interval a part of the reference beam is registered by a photodiode PHD on a storage oscilloscope.

The deformation of the bladders to be inspected is caused by two syringes one of them being used for a prefilling while the other is activated electromechanically giving a continuous piston drive and a continuous increase of the gas pressure inside the bladder while the laser is ignited. From the pulse interval and the rate of pressure increase the pressure difference determinative for the interference process is given. The laser may be ignited by a threshold switch after reaching either a certain bladder volume or a certain bladder pressure.
While the bladder is filled or pressurized three different states are passed: first there is a great volume increase without strain, then a strain of the bladder wall occurs (pressure and volume increase), and finally there is a further pressure increase without a substantial volume increase. The double exposure of the hologram should be done within the second phase when the bladder wall is strained.
The pressure versus volume curves of different bladders deposited in a salt solution or exposed to air for different times show (see Fig.2) that the volume is not suitable for the laser ignition because e.g. at 20 cm^3 some curves are still in the region of volume increase without strain while curve a is already in the region of high pressure increase.

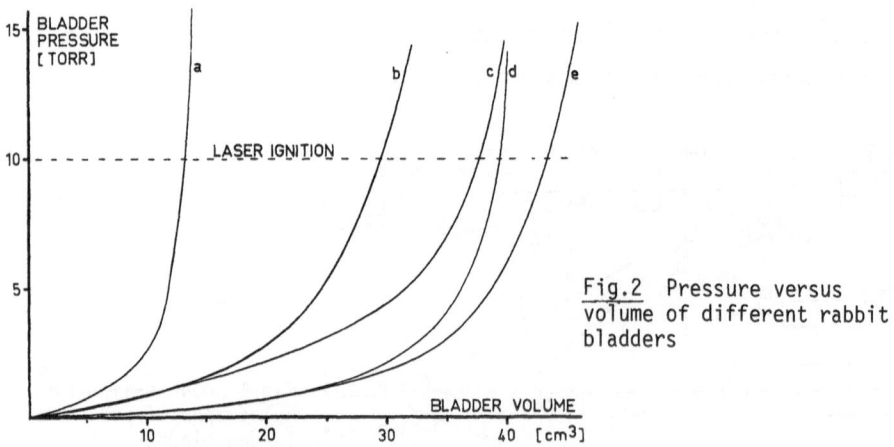

Fig.2 Pressure versus volume of different rabbit bladders

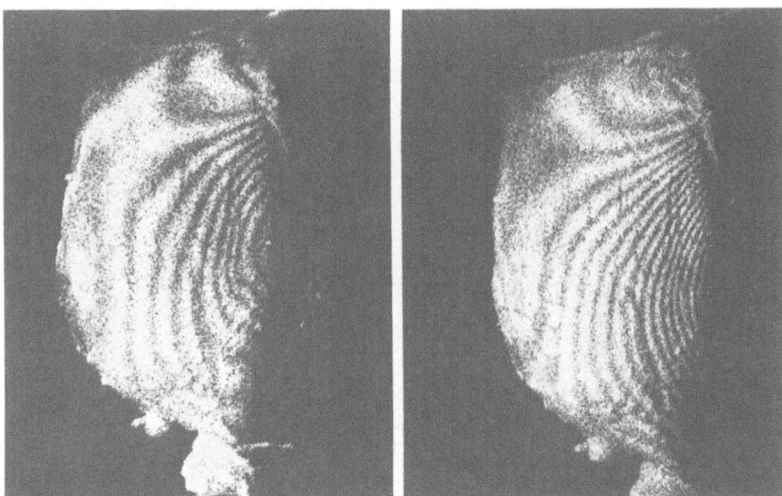

Fig.3   Interferograms of a healthy rabbit bladder

Following this a pressure threshold of 10 torr have been chosen for the
laser ignition. Since these first investigations are rather time consuming all interferograms have been taken with excised bladders.

Previous to the investigations of tumors and bladder lesions [1] the
described arrangement has been tested with healthy bladders concerning
the function of all components. From a lot of interferograms Fig.3 shows
two taken with a filling velocity of 5.5 cm³/s and a pulse interval of
1 ms. In accordance to one another both interferograms show a rather
simple fringe pattern on which local disturbances which could be caused
by natural inhomogeneities of the bladder wall could not be detected.

3.   Endoscopic Holography

Since the later application makes it necessary to take holograms via an
endoscope we made considerations and some experiments concerning this
question, too. To provide a least possible patient's load when more then
one interferogram have to be taken the holographic registration should be
done outside the endoscope.
In contrast to an holoendoscope proposed by HADBAWNIK [2] the hologram
should not be fixed at the top of the endoscope but the optical information should be led out via a normal endoscope-optic for holographic
registration. The Fig.4 shows the concept of this endoholoscope suitable
for holographic image registrations and a direct inspection in parallel.
The illumination is done via a fibre-optic waveguide. The endoscope-optic
projects an intermediate image II near to the hologram H which is illuminated by the reference beam, too.
According to this concept holographic interferograms inside a pressure-
loaded GFRP-tank have been taken using a commercial endoscope (see Fig.5).
Since the tank is a technical object a cw-laser could be used. The
interferograms (Fig.6) show irregular fringes typical to these objects
which have formerly been investigated from the outside in detail [3].

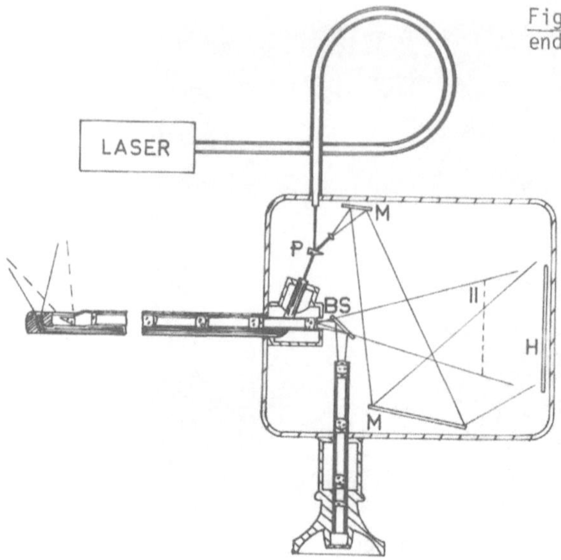

Fig.4 Conceptual draft of an
endoholoscope

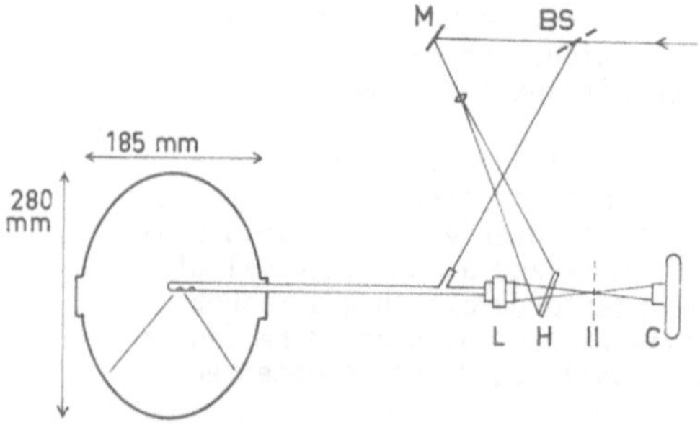

LASER

M
P
BS
II
H
M

Fig.5 Taking holographic
interferograms via an endoscope

185 mm

280
mm

M   BS

L H II C

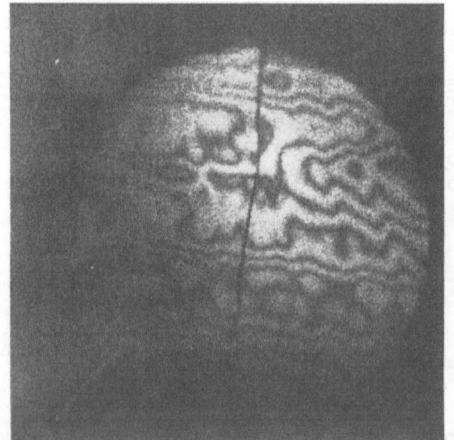

Fig.6 Endoholoscopic interferogram of a GFRP-tank

Beside a lack in illumination in the lower left of the image the quality of the interferogram seems to be adequate. Severe problems with speckles, with wavefront distortions caused by light guidance with fibre waveguides, or with the extended illumination area of some square mm at the fibre optic exit did not occur.

Parts of this work have been sponsered by the Government of the Federal Republic of Germany.

References

1. R. Böwering, A. Hofstetter, K. Grünewald, H. Wachutka: "Interferometric Investigations of the Rabbit Urinary Bladder. II. First Results in the Detection of Tumors and Lesions", in *Holography in Medicine and Biology*, ed. by G. von Bally, Springer Series in Optical Sciences, Vol.18 (Springer, Berlin, Heidelberg, New York 1979) p.152
2. D. Hadbawnik: Dissertation, Stuttgart 1975
3. K. Grünewald, W. Fritzsch: Kunststoff-Rundschau *20*, 593 (1973)

Interferometric Investigations of the Rabbit Urinary Bladder.
II. First Results in the Detection of Tumors and Lesions

R. Böwering and A. Hofstetter

Department of Urology, Hospital
Thalkirchner Straße, D-8000 München, Fed. Rep. of Germany

K. Grünewald and H. Wachutka

Research Department Dornier System GmbH
D-7990 Friedrichshafen, Fed. Rep. of Germany

1. Material and Method

In groups of 5 male giant rabbits with and without bladder carcinomas
(Brown-Pearce transplantation tumors)[1] with a body weight between 1.5
and 1.8 kg, the bladder was exposed under Nembutal anesthesia through a
longitudinal incision in the lower abdomen and fixed to a tripod. The
bladder was filled with air via a 10 French transurethral catheter, and
the double laser impulse elicited at a pressure of 10 torr. The technical
arrangement is explained in the previous paper [2].

After the in vivo measurements, the animal was killed, the bladder removed
and subjected to interferometric holography after various times and pre-
treatments to test the reproducibility of the experimental results in vivo
and in vitro.

In order to improve the contrast of the interference fringes the bladders were
powdered before the holographic interferograms were taken.

In order to be able to compare the interferometric holograms with the actual
extent of the tumor, all the bladders were embedded in celloidin. Photographs
were then taken and a semiquantitative evaluation of the interferograms were
carried out along the lines of intersection which are shown in Fig. 6.

Interferometric holograms were also made of a human cadaver bladder 2 h after
removal from the body. To test the sensitivity of interferometric holography,
exposures were made before and after injection of lubricant jelly into the
bladder wall.

2. Results

Normal rabbit bladder
After preliminary studies interferometric holograms were performed under
different conditions. Figure 1 shows interferograms from healthy rabbit
urinary bladders directly after removal of the bladder and 24 h after
storage in physiological saline at 4⁰ C. All interferograms give evidence
of the reproducibility of the interference pattern: the course of the fringes
shows a regular deformation.

As a measure of deformation the fringe order number was plotted along a
horizontal line across the fringe pattern (Fig.5). However, since the
absolute ordinal number is not known, the actual deformation is given by

Fig.1

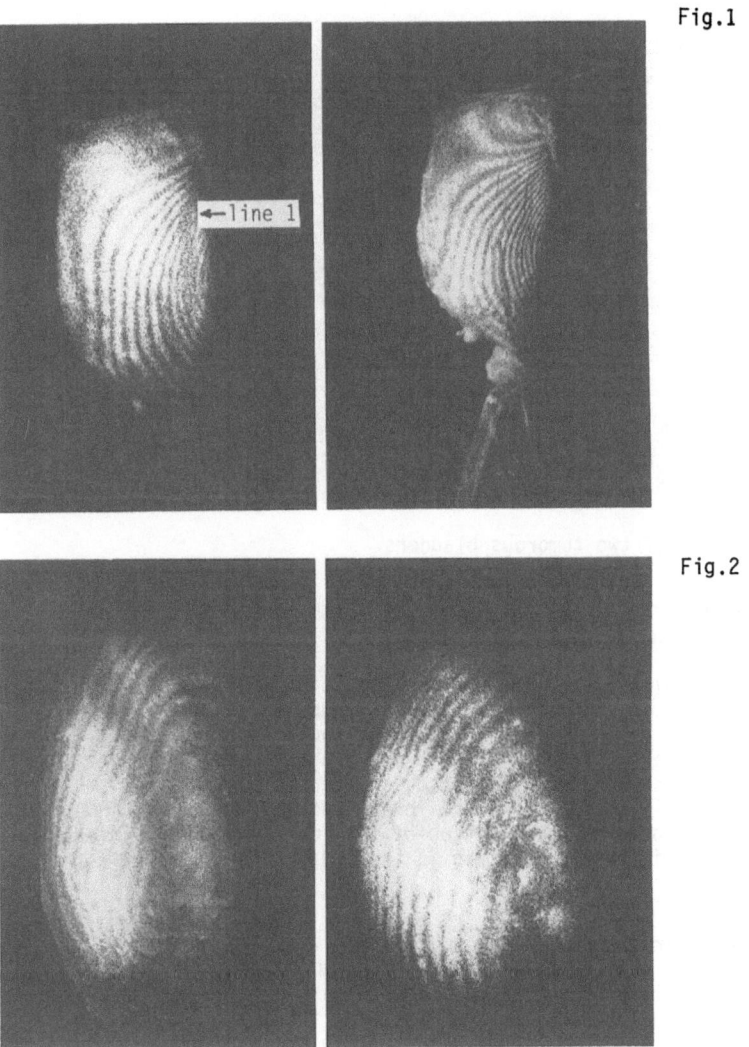

Fig.2

Fig.1 Interferograms of a healthy rabbit urinary bladder

Fig.2 Interferogram of a healthy human bladder before and after
lubricant injection

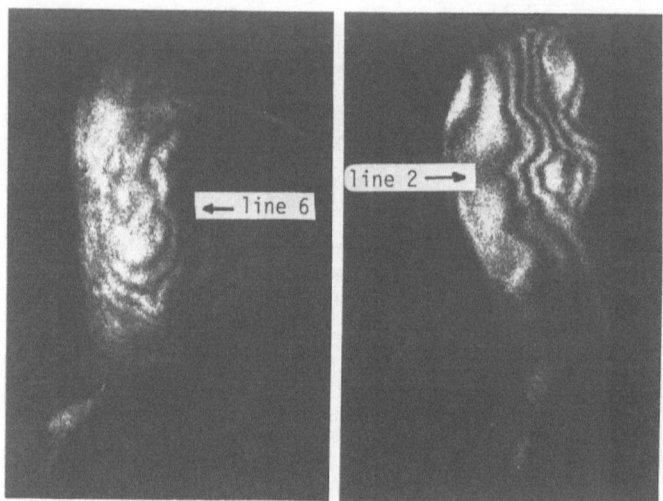

Fig.3 Interferograms of two tumorous bladders

addition of a constant and subsequent multiplication with a geometrical factor as so called sensitivity vector.

The interferometric holograms on the healthy rabbit bladder under various experimental conditions showed interference patterns and regular lines of deformation which were not only largely identical but also comparable to that of the human urinary bladder. Figure 2 shows the interferogram of the human cadaver bladder and the same bladder after injection of 0.5 ml of a lubricant intramurally. Compared to the interferogram, which was taken from the same bladder before the injection of the lubricant, there is a marked bulging of the interference strip in the region of injection.

Tumors
Figure 3 shows interferograms of a tumorous bladder. It can be seen here that the interferences are largely identical. The concentric ring system shows two pronounced disturbances: a small bulge in the lower part of the bladder and a massive disturbance in the middle of the image. As shown by the photograph of this tumorous bladder (Fig.6), these disturbances are identical with the spread of the tumor. In Figure 5, the deformations of this bladder are specified along the lines of intersection. In the region of the tumor, the deformation is markedly disturbed: it is less than in healthy regions because of the sclerosis present here. The good reproducibility in the various experiments in the same bladder is apparent both in comparison with the interferograms and in comparison with the corresponding deformation curves.

In the middle of Figure 4, one interference fringe has a step-like kink (arrow). In "dynamic observation" of the original hologram, this point is still more noticeable because with a head movement the fringes move over the object and seem to stop here. At this point, a punctiform coagulation had been produced with the hot tip of a soldering iron, which resulted in a reduction in the elasticity of the tissue.

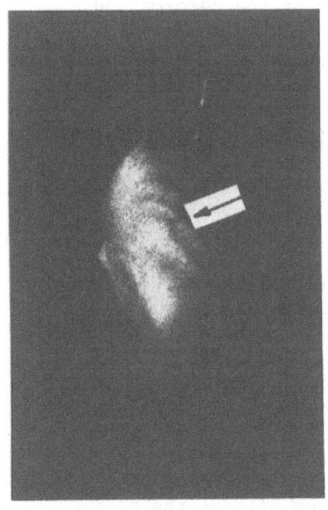

Fig.4 Interferogram of the tumor-free
posterior wall of a bladder with a lesion
(see arrow) after punctiform coagulation

Fig.5 Slope of fringe order across the
interferograms of different bladders (line
1 healthy bladder, line 6 and 2 tumorous
bladders)

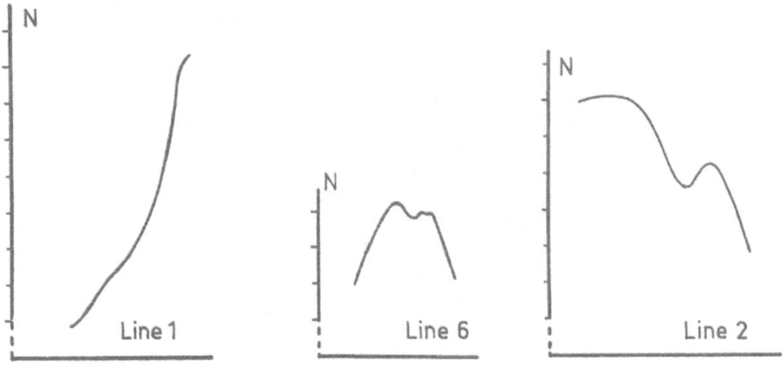

N

Line 1

N

Line 6

N

Line 2

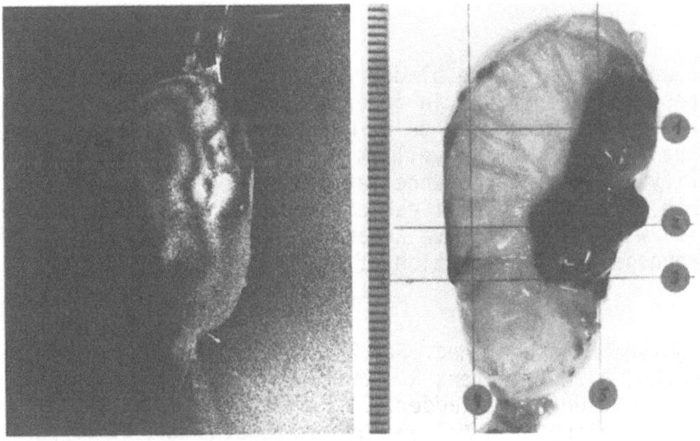

Fig.6 Interferogram of a tumorous bladder and photograph of the same
bladder embedded in celloidin

The bladder embedded in celloidin proves the identity of the deformation in the hologram with the defect in the preparation.

3. Discussion

Our studies proved that interferometric holography is able to demonstrate elasticity changes in biological tissue. In addition interference patterns can be well reproduced under very different experimental conditions, provided that the exposures are taken at a particular stretching pressure of the bladder wall.

It could also be observed that one can distinguish not only between healthy and tumorous altered bladder walls, but that interferometric holography also enabled very exact diagnosis and localization of other elasticity changes for example edema formation caused by thermal lesions, elasticity changes through infiltration of lubricant jelly in the region of the bladder wall. In addition, the quantitative evaluations reveal that the elasticity changes of the tumorous bladder walls cause typical deformation lines. The simple relation was observed that the irregularity of the deformation lines is more intense the nearer the measurement is to the tumor and vice versa. This semi-quantitative evaluation affords the possibility of an exact localization and determination of the extent of the elasticity changes in the bladder wall. The objective of bladder carcinoma diagnostics may be attained; one would be able to perform pinpointed biopsies in the regions of such disturbance fields.

Moreover, it could also be shown that infiltrative growth gives rise to different deformation lines than, e.g. elasticity changes due to (edema development) in the bladder wall when evaluated quantitatively. This means that it should not only be possible with the aid of interferometrical holography to trace areas suspected of containing tumors, but to diagnose these tumors with a high degree of certainty. The designed endoholoscope for the routine application of such a method will permit endoscopic determination of the information on tumorous changes in the bladder wall which we have obtained in the exposed bladder using interferometric holography. Since interferometric holography will be applicable not only to the bladder, but also to other hollow organs in the body and to organic material, we believe that this method will prevail despite the high costs of development in medical diagnostics.

4. Conclusions

Healthy and tumorous changed regions of the bladder exhibit a different stretch behavior in response to alterations in internal pressure. Holographic interferometry enables a distinction to be made between healthy and tumorous regions of the bladder wall as well as regions altered in another way. Reproducibility of the interference patterns under various conditions is well possible. This method applicable to organic tissue and thus to medical diagnostics will be a decisive advance, particularly in the case of the early diagnosis and localization of bladder carcinomas.

References

1. A. Hofstetter, G. Staehler: Fortschr. Med. 95, 346 (1977)
2. K. Grünewald, H. Wachutka, A. Hofstetter, R. Böwering: "Interferometric Investigations of the Rabbit Urinary Bladder. I. Holographic Registration of Bladder Deformations in vitro", in *Holography in Medicine and Biology*, ed. by G. von Bally, Springer Series in Optical Sciences, Vol.18 (Springer, Berlin, Heidelberg, New York 1979) p.147

VIII.   Holography in Dentistry

Holography in Dentistry

H.I. Bjelkhagen

Department of Production Engineering, Royal Institute of Technology
S-100 44 Stockholm 70, Sweden

1. Introduction

The characteristics of laser light i.e. good coherence, monochromaticity, high intensity and parallel nature, provide the
basic conditions for different types of applications within dentistry.

The use of laser light in the field of dental research has
been presented by VAHL [1] and for producing of chemical and
physical alterations by STERN et al [2,3], GORDON [4], SMITH et al
[5], ADRIAN et al [6], SCHEININ et al [7] and KANTOLA [8].

For dental measurement applications the parallel nature and
intensity of the laser light can be used for laser reflection
investigations concerning mobility and deflection, RYDÉN et al
[9, 10].

The coherence and monochromaticity are the characteristics
of laser light mainly used in holography. Hologram interferometry
can be used in different ways for precision interferometric
measurements.

If holography is carried out with a continuous wave laser
(cw laser), the test arrangement must be vibration-free to obtain a hologram of high quality. For this same reason, very rapid
laser pulses ought to be selected for hologram interferometry investigations *in vivo* in order to eliminate the disturbing effects
from inevitable object movements.

2. Hologram interferometry methods

2.1 Contouring

It is possible to produce interference fringes that will represent the intersections of a three-dimensional object by a set of
equally spaced virtually plane surfaces. These *contouring fringes*
reveal the topography of the object just like topography lines
on a map. One way to create these fringes is to double-expose

the hologram, moving the point source that illuminates the object slightly sideways between the two exposures. The reconstructed holographic image reveals the object intersected by surfaces that consist of a set of rotational symmetric hyperboloids, the common foci of which are the two locations of the point of illumination. It is also possible, instead of moving the point of illumination, to change the wavelength of the light slightly between the two exposures, by changing either its frequency or its velocity. If a double-frequency laser is used, the two exposures can be made simultaneously. Again, the holographic image reveals the object intersected by surfaces that consist of a set of rotational symmetric ellipsoids, the common foci of which are the point of illumination and the point of observation.

2.2   Displacement Measurements

Deformation measurements by means of hologram interferometry with double-exposure technique. If the film plate is exposed and the object is then subjected to a load, an exact measure of the deformation produced by the load can be obtained by repeating the exposure on the same plate after deformation. A displacement pattern in the form of alternately bright and dark fringes, the secondary interference pattern, will arise in the image because the wavefronts of the laser light alternately intensify and extinguish each other. This technique is called *double-exposed holography.* The angle of incidence and observation, the wavelength of the light, and the number of interference fringes in the image determine the amount of deformation for the applied load.

Real-time holography. With this technique, it is possible to obtain an interference pattern utilizing a single-exposed plate. After development, the plate is repositioned exactly in the system. The virtual image of the object will interfere with the object itself, and interference fringes will appear on the object if a suitable deforming load is applied. The interference pattern depends on the amount of load on the object.

2.3   Vibration measurements

Time-average holography

It is possible to study the vibration condition of an object using the time-average holography technique, in which a single exposure is made of the object while it is vibrating. If the exposure time is much longer than the time of one oscillation, interference is directly formed between the two outermost positions of the oscillating object. This interference produces a hologram image with fringes that represent the amplitude of vibration at every point on the object surface.

3.   Investigations of dental materials

A Helium-Neon or Argon laser can be used in testing dental materials. Registration of the holograms must be performed in a vibration-free laboratory installation. The holographic interferometry technique utilized in this type of investigation requires two exposures, the second being made after a force is applied to the object.

WICTORIN et al [11] utilized hologram interferometry for investigations of the elastic deformation of defective gold-soldered joints. The interference fringes in double-exposure holography were used to study the elastic deformation of soldered gold structures. The investigation was performed on 10 specimens, defects having been produced artificially in the soldered joints of 5, and on 10 three-unit dental bridges, 5 of which already had superficial defects in the soldered joints. The specimens were loaded, and double-exposure holograms were taken. The results indicated a greater elastic deformation at small loads in the specimens that had defects in the soldered joints.

ALTSCHULER [12,13] made investigations and gave an introduction to dental laser holography. In particular, he has used contouring for making "toothprints" of oral structures. In his opinion, it is possible to modify and combine existing laser, holographic, and computer technology for intraoral use. Individual toothprints may be computerized. As a result, the dentition of patients may be permanently recorded for use in dental examination, forensic odontology, oral diagnosis, security identification procedures, and production of fixed and removable prosthetic appliances, including computer milling of crowns from contour holograms. His report discusses the basic concepts and forms of laser holography, holographic interferometry, contour holographic thermography, and stress holography, as well as their possible dental applications. He also presents techniques developed for holographing such relevant objects as mandibles, skulls, tooth arches, dental casts, prosthetic appliances, and individual teeth.

4. Investigations of prosthodontic appliances

Investigations of the momentary deformations within fixed bridgework at low-level masticatory forces were performed in a simulator arrangement by WEDENDAL et al [14] as a preparation for further clinical experiments. Prosthodontic appliances, especially two types of fixed bridgework, were simulated by means of 7 series of bar elements, cast in a gold alloy. All bars were tested radiographically. Evaluation was made of the positions of defects and areas of reduced density. Elastic deformation studies were made in a bench installation. Double-exposure holographic interferometry was performed, using a CW He-Ne laser. Forces applied to the bar elements were increased between exposures. The effect of force increases up to 1.5 N could be determined by analyzing the resulting interference fringe patterns. Variations in the elastic deformation could be traced to technical reasons, such as the casting technique and the presence of soldered and screw-locked joints. The stiffening effect of internal ivory cylinders simulating teeth prepared as full-crown abutments was proved.

5. Investigations of human hard tissues

The elastic properties of the human hard tissues, especially within the stomatognathic system, have been investigated utilizing mechanical equipment for measurements *in vivo* of the mandibular region. FUCHS *et al* [15,16] studied the deformation of human teeth and related skull bone by means of laser metrology in simulator experiments performed in a laboratory installation. A macerated cranium with complete upper and lower dental arches was used as a test object. The teeth were firmly cemented in their sockets. The masseter muscle force was imitated in a "reverse" way, utilizing an interocclusal soft PVC tube filled with oil. The oil pressure was systematically varied, and the different "masticatory" force levels created were recorded by means of strain gauges. Double-exposed interferograms were registered, utilizing a He-Ne laser. The interference fringe patterns for certain force amplitudes and points of application were studied and verified by means of photographic reconstructions and drawings.

6. Dynamics of human teeth in function

A field of research especially suited to the qualities of laser metrology is the three-dimensional intraalveolar mobility pattern of teeth under physiological and pathophysiological conditions, as well as alterations in this pattern when prosthodontic appliances (periodontally anchored as well as removable) are inserted in the stomatognathic system. Increased knowledge concerning the influence of different types of prosthodontic appliances on the tooth mobility pattern would, for example, be a factor of great importance for the evaluation of prosthetic reconstructions and their long-term prognosis. Since the degree of tooth mobility has for a long time been regarded as an important diagnostic parameter in periodontology, considerably effort has been made to trace the mobility pattern of separate teeth and groups of teeth. It has not been possible to apply most of the measuring methods under physiologic conditions, and few of them have the measuring sensitivity required for recording the motion pattern caused by the very small forces acting on the teeth during certain phases of the physiological function. Such phases occur during mastication (e.g., the degree of force application is very low in the last moment of the chewing sequence) and during the very frequent "empty" or idle movements.

The dynamics of human teeth in function depend on anatomical fundamentals that were already known in the latter years of the 19th century. The teeth are suspended in their sockets by a collagenous fiber system that allows a certain degree of intraalveolar mobility. The arrangement of vascular and nerve bundles completing the periodontal membrane between the tooth and the socket is also of importance for the appreciation of the dynamics of teeth in function. Initial impact on the occlusal surface of a tooth results in a slight compression within parts of the periodontal structures. Vascular bundles thereby act as a kind of hydrodynamic system, having a mild supressing effect on the tooth displacement in the initial phase. Collagenous periodontal fibers have a somewhat wavy internal structure. A slight increase of the acting force causes the fibers to become taut and transmit the force as a pull on the alveolar bone. Increasing force causes an elastic deformation of the alveolar bone, and also involves the elastic root dentine. The point of application, direction, amplitude, and duration of the acting force, as well as the relative lengths of the tooth crown and portion of the root surrounded by the alveolar bone socket, are important factors in this mechanism. Pathologic periodontal damage, interdental splinting, removable appliances, and other factors considerably alter the pattern of tooth mobility.

7. Hologram interferometry *in vivo* utilizing a pulsed laser system

Experimental equipment

Holobeam 651 ruby laser - double-pulsed holographic system. This model is intended for holography. It is constructed to emit a single pulse, or a double pulse with adjustable delay between the pulses. The ruby laser consists of an oscillator and an amplifier. The oscillator or the laser unit consists of a rod made of synthetic ruby surrounded by a helical Xe arc-discharge lamp. A rear mirror and a front mirror with a line selector and a spatial filter create appropriate coherence. A Pockels cell functions as a Q-switch for pulsing the laser. In front of the oscillator unit, there is another ruby rod surrounded by a helical Xe arc-discharge lamp. This ruby rod serves as amplifier.

Subminiature pressure/force sensor. The subminiature pressure sensor used was a Kyowa type PS-10 KA. It consists of a metal cylinder, closed at both ends, with a diameter of 6 mm and a height of 0.6 mm. One of the flat end surfaces is very thin and easily deformed by pressure. Inside it are attached four extremely small strain gauges, which are connected to a bridge circuit. The strain gauges are metal, which in combination with the full bridge arrangement provides the least possible influence of temperature on the measuring results.

In the present investigation, a force sensor was required. In order to transform the sensor from pressure sensor into force sensor, it was placed in a hollow cylinder made of steel, which was closed with a cover about 0.2 mm thick. A central steel rod about 1.3 mm in diameter and about 1 mm high projects perpendicularly to the cover. The masticatory force is applied axially to the tip of the rod, so that the cover is point-loaded.

Clinical experiments

Consideration was given to the utilization of ruby-laser in clinical experiments WEDENDAL *et al* [17,18,19]. If the eye were directly exposed to the laser light, severe retinal burns would result. In these experiments, therefore, the eyes were completely protected by means of black glasses, which had no transmittance for light. A black wooden screen protected the face, leaving the jaw section exposed. Skopyl, 1 ml, was injected about 20 min before the registration in order to inhibit salivation.

For the development of clinical methods, a female patient, 60 years old, was holographed in a series of 10 holographic plates. Her dentition was 47 ... 37 in the lower jaw and 17 ... 25 in the upper. All teeth were very firmly attached to their sockets, and no visible clinical mobility could be recorded. Tooth 24 was chosen to be examined holographically.

The patient´s lips and cheek were retracted by means of a translucent flat acrylic hook. The jaw section to be exposed was painted with gold paint. The cable of the subminiature force sensor was attached to an upper incisor with waxed silk ligature. The sensor was positioned in such a manner that the masticatory force was concentrated in the vertical rod, which was cemented in the distal fossa of 24 by means of acrylic cement. During the cementing, the opposing teeth to be studied were kept simultaneously in contact with the lower metal surface, thus preventing bending and tilting of the sensor during the experiment. The hook as well as the cable to the sensor were located so as not to interfere with the masticatory process. The experiment started with the patient being told to open her mouth wide and then bite down. When the masticatory force, thus applied to the sensor, reached a predetermined level of 2 N, the first laser pulse was triggered. After a delay of 450 μsec, the second pulse was automatically actuated. The double-exposed hologram was then developed and fixed. The force increase and the pulses were simultaneously registered on the oscillascope screen, which was photographed with a Polaroid camera. The force increase between the pulses was compared with the hologram evaluation, and calculations of deformation and tooth mobility could thereby be made. Because of the experimental design used, the force was fairly well defined as regards point of application, direction, amplitude, and duration. One registration is shown in Fig. 1

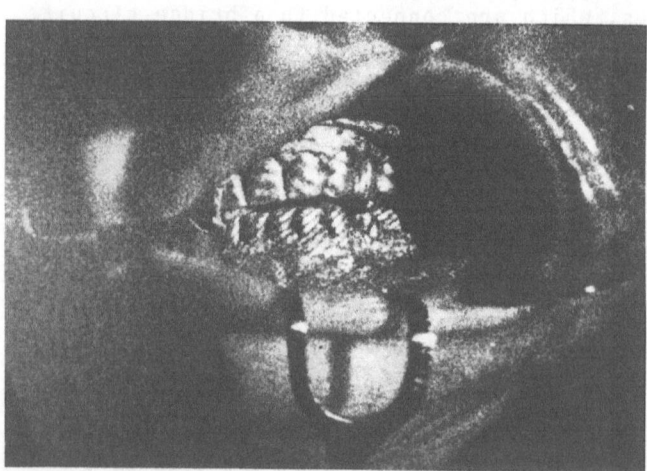

Fig.1 Photographed holographic reconstruction of the exposed jaw section. The subminiature force sensor is positioned between 24 and its opposing teeth. The interference fringes indicate the mobility and deformation caused by the force increase between the laser pulses

8. Speculations for the future

Hologram interferometry provides a noncontact and nondestructive method for measurements of extremely small mobility phases and deformation processes actuated by low-level forces. This technique has formerly been used in industry, e.g., for tests of materials and machined products and for investigations on dental materials and constructions in laboratory installation. This field of research might be of interest in the near future. Simulator tests of different types are not difficult to arrange in the laser laboratory. In this connection, it might be useful to develop computer models to relate forces to deformations in simplified prosthodontic appliances. The sensitivity of laser metrology may be one of the most pertinent new contributions to dental science, if the methods are adequately chosen and properly combined for special purposes. Recent reports exemplify this possibility PRYPUTNIEWICZ [20]. Information on hologram interferometry within dentistry has been presented by WEDENDAL et al [21].

1. Vahl, J., 1971, Der Laser und seine bisherige Anwendung in der Zahnmedizin, Hippokrates 42:488-506.

2. Stern, R. H., Sognnaes, R. F., and Goodman, F., 1966, Laser effect on in vitro enamel permeability and solubility, J. Amer. Dent. Assoc. 73:838-843.

3. Stern, R. H., Vahl, J., and Sognnaes, R.F., 1972, Lased enamel: Ultrastructural observations of pulsed carbon dioxide laser effects. J. Dent. Res. 51:455-460

4. Gordon, T. E., Jr., 1966, Some effects of laser impacts on extracted teeth. J. Dent. Res. 45:372-375

5. Smith, L. D., Burnett, A. P., and Gordon, T. E., Jr., 1972, Laser welding of gold alloys. J. Dent. Res. 51:161-167

6. Adrian, J. C., Bernier, J. L., and Sprague, W. G., 1971, Laser and the dental pulp. J. Amer. Dent. Assoc. 83:113-117

7. Scheinin, A., and Kantola, S., 1969. Laser-induced effects of tooth structure. I. Crater productions with a CO_2-laser. Acta Odontol. Scand. 27:173-179

8. Kantola, S., 1974, Laser-induced effects on tooth structure. Acta Odontol. Scand. 32, Suppl. 63

9. Rydén, H., Bjelkhagen. H., and Söder, P-Ö., 1974, The use of laser beams for measuring tooth mobility and tooth movement. An in vitro study. J. Periodontol. 45:283-288

10. Rydén, H., Bjelkhagen, H., and Söder P-Ö., 1975, The use of laser beams for measuring tooth mobility and tooth movement. A research study of a clinical problem. J. Periodontol 46:421-425

11. Wictorin, L., Bjelkhagen, H., and Abramson, N., 1972, Holographic investigation of the elastic deformation of defective gold solder joints. Acta Odontol. Scand. 30:659-670

12. Altschuler, B. R., 1973, Holodontography: An introduction to dental laser holography. Report SAM-TR-73-4 from USAF School of Aerospace Medicine, Brooks Air Force Base, Texas

13. Altschuler, B. R., and Young, J. M., 1974, Laser holographic stress analysis of removable partial denture connectors, Report from USAF School of Aerospace Medicine, Brooks Air Force Base, Texas

14. Wedendal, P. R., and Bjelkhagen, H., 1974 , Holographic interferometry on the elastic deformation of prosthodontic appliances as simulated by bar elements. Acta Odontol. Scand. 32:189-199

15. Fuchs, P., and Schott, D., 1973, Holographische Interfometrie zur Darstellung von Verformungen des menschlichen Gesichtsschädels, Schweiz. Monatsschr. Zahnheil. 83:1468-1482

16. Fuchs, P., and Schott, D., 1973, The application of holography to the measurement of deformation of human facial bones, Dtsch. Zahnärztl. Z. 28, 90

17. Wedendal, P.R., and Bjelkhagen, H., 1974 , Dental holographic interferometry in vivo utilizing a ruby laser system. I. Introduction and development of methods for precision measurements on the functional dynamics of human teeth and prosthodontic appliances. Acta Odontol. Scand. 32:131-145

18. Wedendal, P. R., and Bjelkhagen, H., 1974 , Dental holographic interferometry in vivo utilizing a ruby laser system. II. Clinical applications. Acta Odontol. Scand. 32:345-356

19. Wedendal, P. R., and Bjelkhagen, H., 1974 , Dynamics of human teeth in function by means of double pulsed holography: An experimental investigation. Appl. Opt. 13:2481-2485

20. Pryputniewicz, R. J., 1979, Holographic determination of rigid body motions and applications to orthodontics. Appl Opt. In the print.

21. Wedendal, P., and Bjelkhagen, H., 1977, in Laser applications in medicine and biology, Volume 3, edited by Wolbarsht, M. L., (Plenum Press, New York), chapter 3:221-287

Holographic Evaluation of the Dimensional Stability of Elastic Impression Silicones Used in Dentistry

J.P. Goedgebuer, M. Spajer, and J.Ch. Vienot

Laboratoire de Physique Générale et Optique, associé au CNRS
"Holographie et Traitement Optique des Signaux"
Université de Franche-Comté, F-25030 Besançon Cedex, France

SUMMARY

The paper deals with a problem of deformation applied to dentistry when making imprints with precision impression materials such as silicone elastomers. A study of their dimensional variations has been carried out through (i) holographic interferometry in laser light, (ii) interferometry in white light based upon the channelled spectrum phenomenon. Both high and low-bodied elastomers have been tested. The results indicate the influence of the thickness of the sample on the speed of retraction during the polymerization of the material. Furthermore, external parameters such as the temperature and the hygrometric degree have to be taken into account when making accurate imprints.

1. - INTRODUCTION

Silicone elastomers are widely used in dentistry for making accurate imprints of teeth. The operation consists in inserting in the mouth the silicone-paste mixed with a hardening liquid. The paste becomes elastic after a setting time corresponding to the polymerization of the material. The imprint so obtained is used as a negative matrix mould allowing the prosthesis itself to be modelled. Final success depends on the accuracy of the moulding process, that generally ranges between 40 and 80 μm. Unfortunately the silicone-rubber moulds are not stable in time and suffer from dimensional variations depending on several parameters [1], such as the chemical constitution of the paste, the amount of hardening liquid, the moulding technique the conditions of impression removal, etc... The techniques of holographic interferometry and channelled spectra are reviewed and discussed as methods for measuring these variations.

2. - DOUBLE-EXPOSURE HOLOGRAPHY [2,3]

Fig. 1 a, b and c are photographs of double-exposure interferograms obtained from a light-bodied sample of prismatic thickness profile (Fig. 1 d). The sample is moulded on a glass plate, after mixing the silicone paste with the hardener. The prismatic profile offers the advantage on the one hand of determining the fringe order number without any ambiguity since the surface displacement is zero near the edge of the prism, and on the other, to evaluate the dimensional stability as a function of the thickness in one

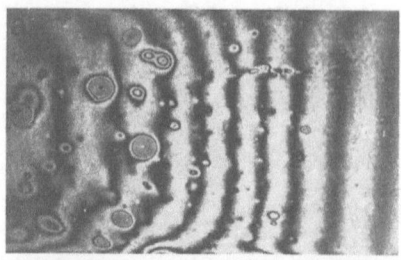

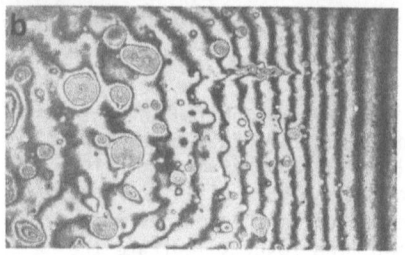

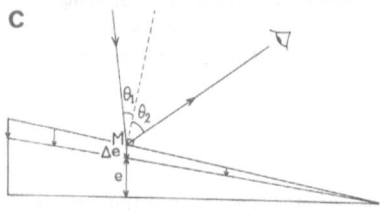

Fig. 1 : *(a) and (b)* : *double-exposure*
interferograms recorded at different
times   after the moulding process.
the sample is a light-bodied elastomer
prismatic-shaped ("Blue Xantopren" from
Bayer) ; e denotes the thickness of the
material along its prismatic profile (c)

unique experiment. In Fig. 1a
the fringes code a relative
decrease of thickness of about
10^{-3} over a period of 45 mn
after the setting time. The
circular fringe patterns in
the interferograms denote
thickness irregularities due
to air bubbles collapsing near
the surface when moulding the
sample. As mentioned previously
the interferograms discussed
so far concern a light-bodied
silicone elastomer moulded on
a glass plate - that explains
that no speckle is observed
in the interferogram, the
surface under test being
perfectly polished. In fact,
the working conditions depend
on the chemical constitution
of the paste, especially its
viscosity. As an example,
Fig. 2 is an interferogram
obtained with a high-bodied
silicone elastomer moulded
in the same conditions. The
surface under test is now
strongly diffusing and a
speckle pattern spoils the
interference fringes in the
interferogram. Moreover, experiments [5,6] show that the
interference fringes may
vanish completely. Such a
situation is due to a decorrelation of the speckle

patterns between the two exposures of the photographic plate [3] ; it points
out that microscopic deformations of the surface appear simultaneously with
a retraction of thickness. As a consequence, measurements have to be carried
out recording several successive double exposure holograms sufficiently
close in time. Last, heavy-bodied silicone elastomers are difficult to mix
homogeneously with the hardener and local variations in the mixture introduce a granular structure in the fringes - that makes difficult the measurements.

A palliative consists in using a small mirror on the surface
under test. This method gives access to mean variations of thickness and
removes any difficulty linked with occasional air bubbles, microscopic

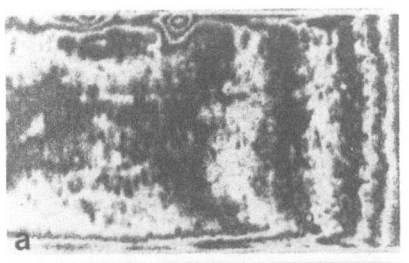

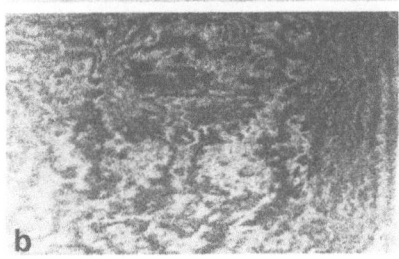

Fig. 2 : *Speckled interferograms of a high-bodied silicone elastomer ("Optosil" of Bayer) moulded on a glass plate. Fig. 2(a) evinces the granular structure of the fringes. In (b) the fringe visibility is altered, pointing out a decorrelation of the microscopic structure of the surface between the two exposures of that inter-ferogram.*

surface deformations or a non homogeneous mixture. This solution has been tested through the channelled spectrum technique [4], in the frame of a work dealing with the so-called "space-time optics" [6].

3. - RECORDING OF CHANNELLED SPECTRUM

Fig. 3 illustrates the principle of operation and Fig. 4 represents the set-up. The sample is inserted along one arm of a two-beam interferometer illuminated in white light. The path-difference introduced by the object is $\Delta(x)$. The spectral analysis of the light emerging from the interferometer yields a channelled spectrum [4]. The power-spectrum $B'(\sigma)$ along a σ - axis ($\sigma = 1/\lambda$: wave-number) in the spectrogram is related to the path-difference $\Delta(x)$ by :

$$B'(\sigma,x) = B(\sigma)\cdot \{1 + \cos 2\pi\sigma\Delta(x)\} \qquad (1)$$

where $B(\sigma)$ is the power-spectrum of the white light source. The measurements are obtained by computing the number N of fringes channelling a spectral bandwidth $|\sigma_2 - \sigma_1|$, since, from (1):

$$N = |\sigma_2 - \sigma_1| \cdot \Delta \qquad (2)$$

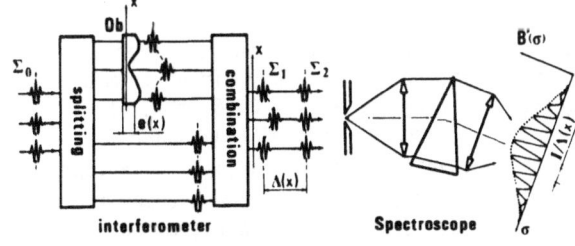

Fig. 3 : *The channelled spectrum technique for measuring an optical delay Δ, $\sum_1$ and $\sum_2$ being two delayed wave-fronts of white light after passing through the interferometer. Determination of the fringe spacing in the spectrogram leads to Δ (Ref. 4 and 7).*

Fig. 4 : *Gauge based on the channelled spectrum phenomenon* [8];
S : white-light source ; M : Michelson interferometer ;
G : grating ; C : camera.

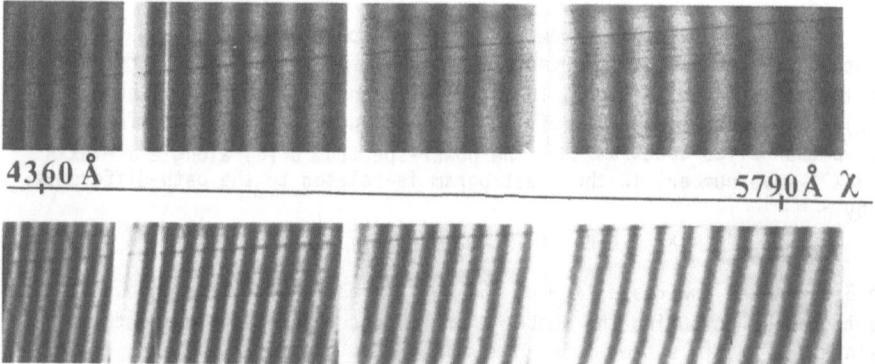

4360 Å   5790 Å λ

Fig.5 : Channelled spectra produced by a silicone elastomer during its polymerization. The interval of time between the two exposures is about 50 hours.

An example gives the orders of magnitude. Fig. 5 represents two spectrograms
recorded during the polymerization of a silicone sample. They differ from a
number $\delta N = 10$ of fringes in a spectral interval $|\sigma_2 - \sigma_1| = 0.56 \ \mu m^{-1}$. It
means, from (2), that the retraction of thickness between the two expo-
sures corresponds to a variation of the path-difference of $\delta N / |\sigma_2 - \sigma_1| \approx 18 \mu m$.

Compared to holographic interferometry, this method overcomes the
uncertainty of $k\lambda$ in the measurement of a path-difference. Therefore, no
thickness profile of any kind (e.g. prismatic type) is required and the measurements are carried out with materials of any profile. Moreover the tech-

nique seems non sensible to speckle phonomena and therefore appears quite suitable for studying high-bodied elastomers, especially as microscopic deformations of the surface structure occur. However, white light sources of high luminance are required.

4. - RESULTS

The curves in fig. 6 are typical examples showing the relative variations of the thickness during the polymerization reaction that may last for more than 48 hours. From our results in publication elsewhere [7] some conclusions can be summarized :
- the dimensional stability is better as the quantity of hardener mixed with the silicone base is smaller;
- the material is non stable during the first hours after the moulding process, corresponding to a first stage of polymerization;
- the stability increases as the thickness of the sample decreases;
- many external parameters intervene in the dimensional stability of the silicone elastomers (temperature, hygrometric degree, sample at open air, etc...).

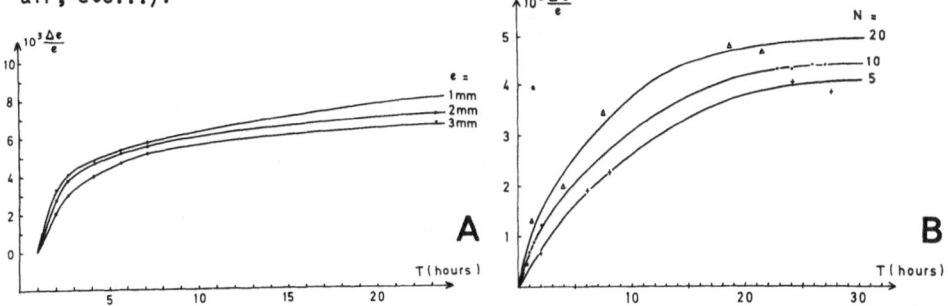

Fig.6 : Relative variations $\Delta e/e$ of thickness of "Blue Xantopren" in (a) and "Optosil" in (b). Complete results are given in ref.6

1. C. Pirel, Thèse de 3e cycle, Université Claude Bernard, Lyon 1, UER des Sciences Odontologiques.
2. K.A. Stetson, R.L. Powell, J.O.S.A., 55 (1965), p. 1604
3. J.-Ch. Viénot, C. Froehly, J. Monneret, J. Pasteur, Symp. on the Engineering of Holography, Glasgow (sept. 68), Proc. Ed. by E.R. Robertson, Camb. Un. Press, London 1970, p. 133-150
4. e.g. H. Bouasse, Z. Carrière, "Interferences" (Delagrave Ed. 1923)
5. M. Blandin, C. Durou, H. Soulet, SPIE vol. 136, 1st European Congress on Optics Applied to Metrology (Strasbourg 1977), pp. 130-134
6. J.-Ch. Viénot, J.P. Goedgebuer, A. Lacourt, Applied Optics, 16,2, 1977, 1977, pp. 454-461
7. M. Spajer, J.P. Goedgebuer, in publication in Optics and Laser Technology (April 1979)
8. J.P. Goedgebuer, A. Lacourt, M. Guignard, Optics and Laser Technology, August 1978, pp. 193-196.

Measurements of Deformations of Teeth and Mandibles due to Occlusal Forces

T. Matsumoto[1], T. Fujita[1], R. Nagata[2], T. Sugimura[3], and Y. Kakudo[3]

[1]Osaka Prefectural Technical College, Dept. of Mechanical Engineering
Saiwaicho, Neyagawa, Osaka, Japan
[2]University of Osaka Prefecture, Dept. of Mechanical Engineering
College of Engineering, Mozu, Sakai, Osaka, Japan
[3]Osaka Dental University, Dept. of Oral Physiology
Kyobashi, Higashiku, Osaka, Japan

Abstract

When occlusal or masticatory forces were applied to teeth and
mandibles, three dimensional (3-D) displacements of their deformations were measured quantitatively by double-exposure holographic
interferometry. The mandibles were extracted from adult dogs. As
measuring places, buccal surfaces of the teeth and mandibles were
selected. When vertical load was applied to the first retromolar
(M1), a tiltlike deformation was observed. By this the tooth was
displaced to the buccal direction. Experimental observation of a
sagittal section shows that deformations of the tooth are larger
than that of the corresponding mandible. When a lateral load acted
on M1, the buccal surface of M1 was displaced into the upper direction of the tooth axis from the dental socket. In these experiments, two kinds of tensile forces, 100 gm and 200 gm were applied
to M1. When the buccal surface of the mandible in the neighbourhood
of M1 was partly scraped, the same surface of the tooth was displaced further to the upper direction of the tooth axis. The mandible in the neighbourhood of M1 was displaced in the same direction.
These results show that the periodontal membrane plays the role of
a buffer and will be valuable to considerations of deformations of
human teeth and periodontal diseases.

1. Introduction

Quantitative measurements of deformations of teeth and mandibles
due to occlusal or masticatory forces are important in oral physiology, prosthetic dentistry, and orthodontics. In many reports
mechanical methods habe been presented [1,2]. Recently, P.R. WEDEN-
DAL and H.I. BJELKHAGEN have reported on investigations of the
functional dynamics of human teeth by double-pulsed holography
to develop a noncontact and nondestructive method [3,4].

This paper describes results of quantitative measurements of
3-D displacements of teeth and mandibles under various forces. It
is known that alveolar resorption is caused by alveolar blennor-
rhea or malocclusion. In such cases, we are interested in measuring
deformations of teeth and mandibles subjected to lateral load. When
a mandible was scraped to one-third or two-thirds of its height from
the alveolar crest, deformations of buccal surfaces of the teeth

and mandible were measured. The forces were applied in vertical
and lateral directions, respectively. The paper consists of three
parts: measurements of deformations of teeth and mandible due to
vertical load applied to the fossa of M1, due to lateral load
applied to the lingual side of M1 with and without the mandible
being scraped to one-third and two-thirds of its height.

2. Experiments

2.1 Vertical load applied to the fossa of M1

Mandibles together with the teeth of adult dogs were extracted
under sodium pentobarbital anesthesia (Nembutal). The gingiva was
removed. Figure 1 shows the mandible and teeth used in this experiment. The tooth axis of M1 was placed perpendicular to the
upper plane of a super hard plaster. Then the mandibular base was
cemented into the plaster with alpha cyanoacrylate adhesive (Aron
Alpha). As it is well known in oral physiology the maximum occlusal force acts on M1, the vertical load was applied to this tooth.
The load was applied by a simple apparatus using a weight. The
optical system to make the double-exposure holographic interferograms is shown in Fig. 2. The light beam is divided into a transmitted beam and a reflected beam by a glass plate.

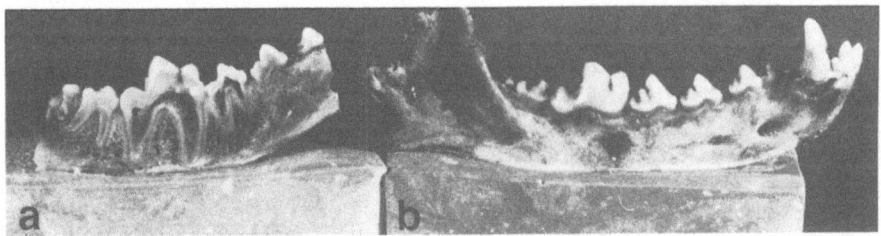

Fig. 1 Specimens used in the case of vertical load applied to the
fossa of the first retromolar (M1). (a) and (b) show the case of
sagittal section and *in situ*, respectively.

The transmitted beam illuminates normally the object. The mandible is covered with white paint to scatter the incident light
effectively. The reference and object beams illuminate the photographic plate. After the object is subjected to vertical load,
the second exposure is recorded on the same plate. In the reconstruction process, the reference beam illuminates the hologram.
The reconstructed image can be recorded with a camera. Figure 3
shows one of the experimental results from the *in situ* case.
Figure 3(a) presents the displacement distribution caused by small
vertical load, Figure 3(b) and Figure 3(c) show patterns caused by
increasingly large vertical loads.
In order to evaluate the deformations of the supporting structures, e.g. periodontal membrane, cancellous bone, and compact
bone, the lingual side of the mandible was removed with a trimmer.
Figure 4 shows a reconstructed image. We can see a shift of the
interference fringes at the boundary between dental root and compact bone. It is considered that this is an effect caused by the
periodontal membrane.

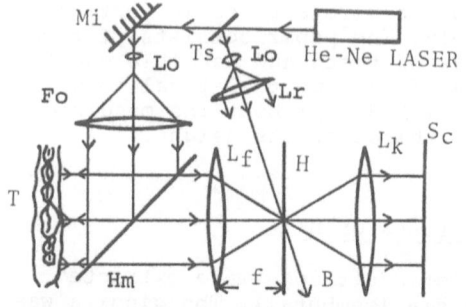

Fig.2 Optical system for recording and reconstructing double-exposure holograms. Hm: beam splittering mirror, Lf: Fraunhofer lens, H: hologram, B: reference wave, Mi: mirror, Rs: glass plate, Lr: collimating lens, Fo: illuminating lens, Lk: imaging lens, Sc: screen, T: teeth and mandible

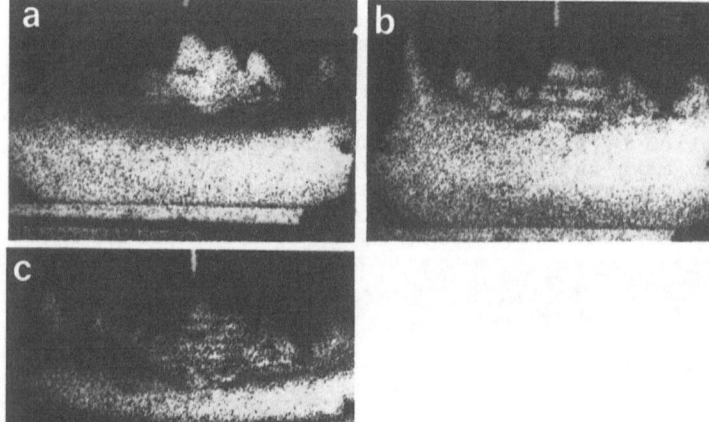

Fig. 3a-c Reconstructed images obtained at three different vertical loads

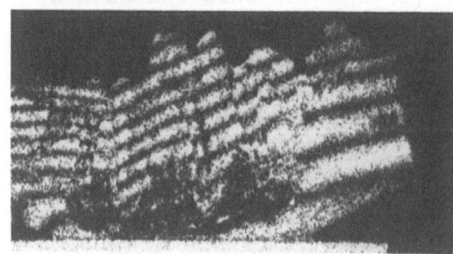

Fig. 4 Reconstructed image obtained from a sagittal section

2.2 Lateral load applied to the lingual side of M1

Figure 5 shows schematic diagrams of a specimen. Lateral load acts on the lingual side of M1. Nine points on the upper plane of super hard plaster were selected as shown in Fig. 5(b) to apply the load to M1 from different directions.

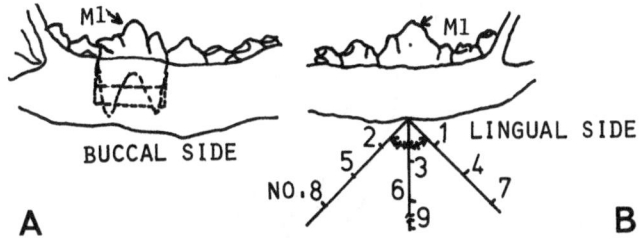

Fig. 5 Schematic diagrams of a specimen as used in the case when lateral load is applied. (a) and (b) show buccal and lingual side view, respectively. Points from No. 1 to No. 9 denote loading points on the upper plane of super hard plaster. Dotted lines outline the parts of the mandible, which are scraped from the alveolar crest as described in 2.3.

An illustration how the load is applied, is given in Fig. 6. As shown in this figure, one end of a wire is fixed to a point on M1 with a hook. The wire is supported with two small steel ball bearings to make friction small. The other end of the wire is fixed to a weight.

The optical system to measure the 3-D displacements is shown in Fig. 7. I and O_3 are in x-z plane. Three Fraunhofer holograms are recorded with the optical systems as shown in Fig. 2 for three observing directions O_j (j=1,2,3). Measuring accuracy of 3-D displacements is described in [5]. For this optical system, the equations to obtain 3-D displacements U_x, U_y and U_z are given as follows:

$$U_x = 0.0511(N_1+N_2)+0.2930N_3$$
$$U_y = 0.5167(N_1-N_2)$$
$$U_z = -0.2978(N_1+N_2)+0.5995N_3$$

$$(1)$$

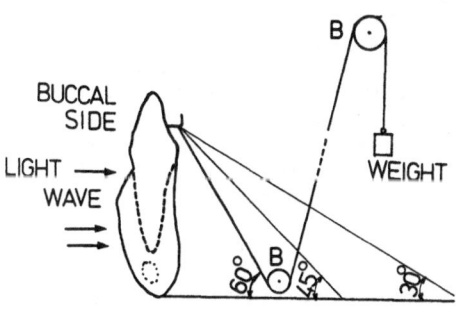

Fig. 6 Schematic diagram to illustrate the tensile force applied to the first retromolor, B: ball bearing.

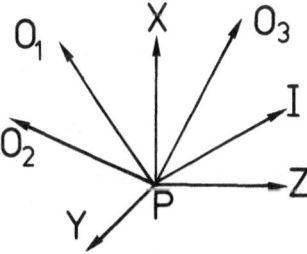

Fig. 7 Coordinate system to measure 3-D displacements. The origin P is an arbitrary point on the object. I and O_j (j=1,2,3) are unit vectors of illuminating and observing directions, respectively.

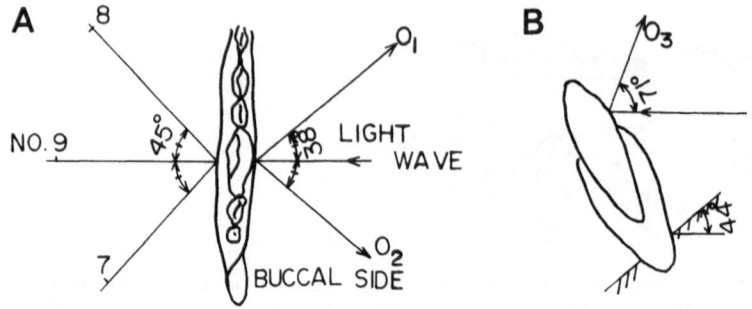

Fig. 8 Schematic diagram to illustrate three observing directions O_j (j=1, 2, 3). (a) and (b) show plane and side view , respectively.

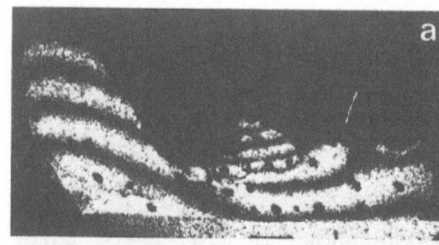
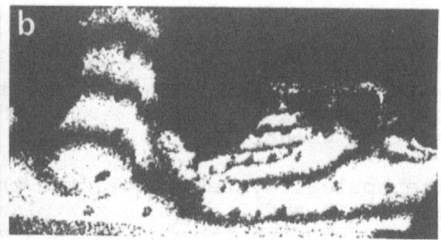

Fig.9 Deformations of the tooth M1 and the mandible due to a tensile force of 200 gm, applied from the point No. 9. (a), (b), and (c) show reconstructed images obtained from O_j.

where N_j (j=1, 2, 3) are fringe order numbers obtained from each hologram. U_x, U_y and U_z denote the displacement components along the coordinate axes. In this experiment, the measuring accuracy of displacement was 0.2 μm in absolute error and 6% in relative error [5]. Loading, observing and illuminating directions for the tooth and mandible are shown in Fig. 8. When we know N_j from three reconstructed images, 3-D displacements can be calculated by (1). In Fig. 9, deformations of the teeth and mandible are shown for a tensile force of 200 gm, applied from the point No. 9.

2.3 Mandible scraped partly from the alveolar crest

Specimens were scraped to one-third and two-thirds from alveolar crest as shown by dotted lines in Fig. 5(a). Fig. 10 shows reconstructed images obtained under these experimental conditions. As we can observe immediately interference fringes are concentrated on M1. This tooth has a larger deformation than the other teeth and the mandible.

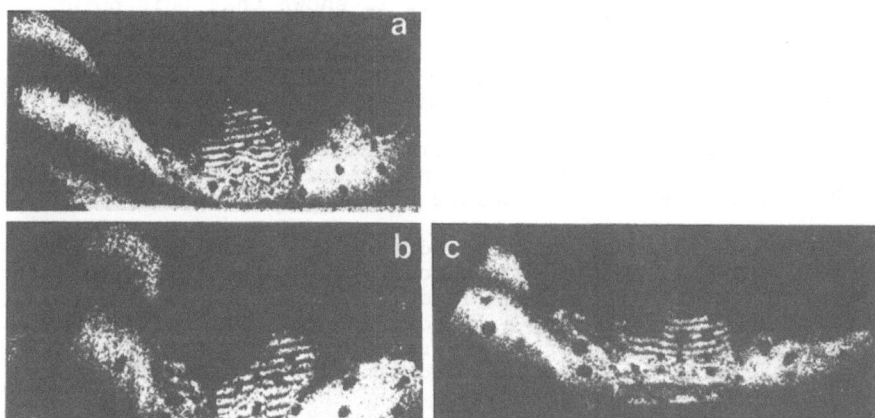

Fig. 10 Deformations of the tooth M1 and the mandible due to a tensile force of 100 gm. The mandible was scraped to two-thirds of its height. (a), (b) and (c) show reconstructed images obtained from three observing directions O_j. The force was applied from point No. 9.

3. Considerations and Conclusion

Schematic diagrams in Fig. 11 show displacement vectors in two points on M1 for three experimental conditions. In the figure, A, B, and C mean the case *in situ*, of one-third removal and of two-thirds removal, respectively. Numbers 1 and 2 define the points on the cusp and in the neighbourhood of the alveolar crest

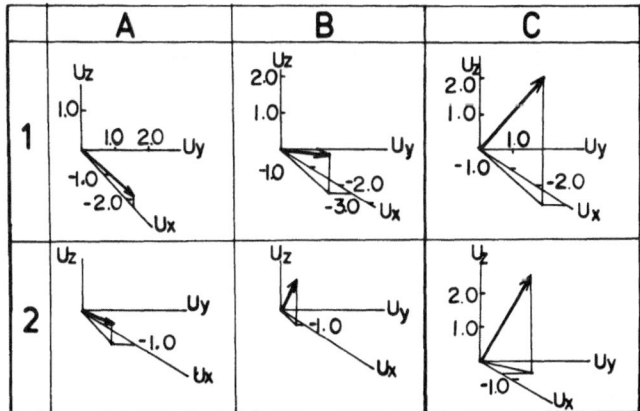

Fig. 11 Measuring results of 3-D displacements due to the force applied from the point No. 9. A, B and C denote the mandible being *in situ*, scraped to one-third and two-thirds of its height, respectively. Numbers 1 and 2 define the points on the cusp and in the neighbourhood of the alveolar crest of M1, respectively.

of M1, respectively. In the case C, it is shown that this point is displaced 3.6μm in upper direction, 0.7μm in distal direction and 3.0μm in lingual direction. From these results, it can be understood clearly that the tooth moves into upper direction with increasing amount of removal of the mandible.

When the force acted on M1 from other places on the super hard plaster, the results were the same in direction of displacement vector but different in the absolute value.

Following conclusions are obtained from the measuring results. When a vertical load acts on M1, we can observe interference fringes parallel with the upper plane of the plaster on the teeth and mandible. This shows that a uniform displacement direction of the teeth and mandible occurs for vertical load. From experimental results in the case of a sagittal section, we can find that the deformations of the teeth are larger than that of the mandible and that the periodontal membrane plays the role of a buffer. When a lateral load acts on M1, two conclusions are obtained. When the mandible is *in situ*, we can find that displacements of the tooth are larger than that of the mandible. This shows that teeth are rather movable when lateral load is applied. Removing a part from the top of the alveolar bone, teeth become very movable and the tooth M1 is deformed into upper direction from the alveolar socket.

These conclusions were obtained from experiments on dog teeth, but they may also be interesting for considerations on deformations of human teeth.

References

1. H.R. Mühlemann, Oral Surg. <u>44</u>, 1220 (1951).
2. D.C.A. Picton, Arch. Oral. Biol. <u>9</u>, 565 (1964).
3. L. Wictorin, H. Bjelkhagen and N. Abramson, Acta Odont. Scand. <u>30</u>, 659 (1972).
4. P.R. Wedendal and H.I. Bjelkhagen, Appl. Opt. <u>13</u>, 2481 (1974).
5. T. Matsumoto, K. Iwata and R. Nagata, Appl. Opt. <u>12</u>, 961 (1973)

Strain Distribution in the Facial Skeleton Arising from Orthodontic Appliance Activity

P. Pavlin, D. Vukicević, and Z. Rajić
Polyclinic of Dentistry, Zagreb, Perkovceva 3
Institute of Physics and Faculty of Dentistry of the University of Zagreb
Yu-Zagreb, Yugoslavia

The methods of biomechanical research and the principles
arising from its results have been successfully applied to orthodontics. From the wide range of problems that biomechanics deals
with, orthodontics primarily utilizes the results of research in
investigations of bone properties of the dento-facial complex
and its reaction to physiological and non-physiological forces.
The interpretation of the way strain acts on the above mentioned
bones is an extremely difficult problem because of the complex
anatomical structure and anisotropic properties of the bones.
In physical terms, bone acts as a viscoelastic system, so that its
deformation depends not only on the magnitude and direction of
the applied force but also on the time during which the force
acts /1/.
   Although the problem is one that has been treated by many
authors, the existing knowledge of the mechanical properties
and the reactions of the skull bones is rather insufficient.
The first significant results came from BENNINGHOFF's work /2/,
which proved his hypothesis of the trajectory of the forces
that dissipate the masticatory pressure, followed by the contributions of PAUWELS /3/, who investigated the architectural principles of the locomotor system in regard to stresses of bones.
Attempting to explain the strain distribution on the cranial
skeleton, one of the most useful methods is the application of
strain-gauges /4/, especially when used with respect to the
structural analysis of bones in the masticatory system. In
addition to this tensometric method, recent investigations of
bone strain have used the brittle enamel method as well as photo-
elastic measurement /5/. The latter method proved the validity
of criticism directed against the method of isotropic photo-
elasticimetry in combination with tensometry /6/.
   Over the last decade the advantages of holographic interfer-
metry, as a non-destructive method that permits experiments on
anatomical specimens, have been utilized by many authors dealing
with biomechanical properties of the dento-facial complex /7,8,9,
10,11,12,13/.

Materials and Methods

The purpose of the investigation described is to determine the
strain distribution on the upper jaw and the dental arch skeleton, arising from the action of removable orthodontic appliances
and to compare the results obtained with those of other investi-

gations of the strain distribution in this part of the skeleton during the masticatory function.

The experiments were carried out on an anatomical specimen of the human skull, taken from a fresh cadaver which was macerated immediately before the start of the experiment, so as to preserve the mechanical properties of the natural bone as much as possible.

The specimen had a perfectly preserved processus alveolaris and all teeth aligned in the regular form in the dental arch. In order to obtain a suitable anatomically formed support for placing the appliance the soft pallate tissue was reconstructed to its approximate thickness. The material used in the reconstruction was a mixture of hard and elastic auto-polymerized acrylics, so that it was possible to obtain resilience close to that found in vivo. The destruction of the periodental tissue during maceration made it necessary to firmly fix the teeth in their respective alveolae. This was achieved using the above mentioned combination of auto-polymerized acrylics. Although the teeth were so fixed in their alveolae, the force was nevertheless elastically transmitted from them to the alveolar bone.

For the jaw, two active acrylic plates were made by the usual laboratory procedure; two arrow clasps provide the anchorage for the left side, and the Adams and arrow pin clasp for the right. The clasps were built into both appliances identically. Both plates were divided and driven apart by screws. One of the plates was symmetrically divided, the other asymmetrically by cutting it 3 mm to the right from the sagittal axis. This choice was made because the active plate is the most frequently used orthodontic appliance, and its active elements are combined in various different ways in orthodontic treatment. On the other hand, jaw expansion is often the first step in obtaining normal relations of the jaws, and we presumed that the determination of bone strain distribution during this type of therapy would be very interesting from the orthodontic point of view.

The skull was fixed within a specially constructed frame at four points of the cranium. The arrangement for holography was set up so that separate holograms of the left and right maxilla, resp., could be recorded simultaneously as shown in Fig. 1.

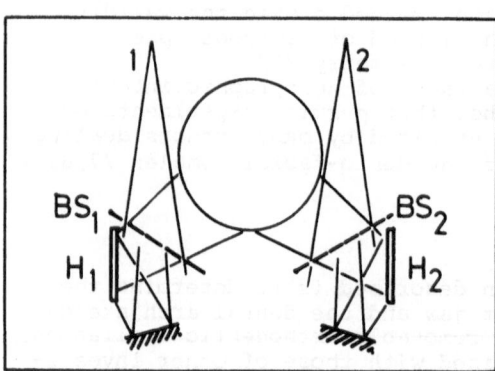

Fig. 1. Optical arrangement

Since the stress-loading due to the orthodontic appliance has a significant component in the lateral direction of the horizontal plane, and because the comparative measurements had to be taken simultaneously from both sides in a relatively restricted area whose diameter was not greater than 10 cm, interferograms were made using beam splitter plates. The direction of the illumination and the direction of observation were identical, which corresponds to an object position on the major axis of the holodiagram /14/ and a sensitivity factor of k=1. Initially, a collimated beam was used but the system of confocal ellipsoids on the hologram was later reduced to a system of parallel planes in order to eliminate systematic errors /15/. After the first set of interferograms was recorded, it was clear that the systematic error in using divergent beams is negligable compared to the advantages arising from the illumination of a greater surface area of the object.

Experiments and Results

When the appliance is fixed to the jaw it makes contact with the palate and the palatal side of the teeth. By turning a screw, the plates are forced apart and act on the teeth and palate with a laterally directed force, approximately perpendicular to the sagittal plane. The degree of rotation of the screw determines the movement of both plates in the lateral direction. A rotation of 90 degrees corresponds to a movement of the plates of 0.2 mm, i.e. 0.1 mm to each side. The forces acting in such a case vary individually and depend on the tissue response to the pressure of the plates. The response is a function of the mechanical properties of the tissue being put under pressure, and varies according to the individual morpho-functional properties of the dento-facial complex /17/, the age and sex.

The first set of holographic double-exposure interferograms was recorded using a symmetrically divided appliance without pre-stressing. This means that the first exposure was made with the screw firmly fixed, and the second one after turning the screw. In the first interferogram the rotation of the screw was 15 degrees, which corresponds to a movement of the plates of 0.017 mm to each side. Each of the following interferograms was taken at a further 15 degrees turn of the screw for the second exposure, thus with each turn increasing the displacements of the plates by 0.017 mm to each side.

The second series was recorded with pre-stress, so that the zero aperture was 0.2 mm greater (90 degree turn) than that in the first series. For each interferogram the added aperture was 0.033 mm (which corresponds to a 30 degree turn). The third series of interferograms was recorded using an asymmetrically divided plate, with and without pre-stress.

All the holo-interferograms show many fringes which can be perfectly traced on the maxilla (Fig. 2). It was therefore decided to analyse also the deformations of the facial skeleton, and direct the investigations primarily to the comparison of these deformations with changes occuring during the action of masticatory forces. We also found that our experiments using the asymmetrical appliance were not satisfactory. The results for the two sides of the jaw were not comparable because there was an

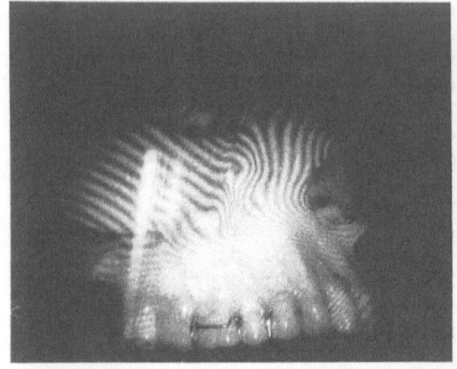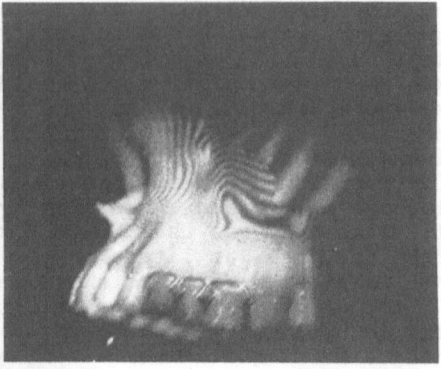

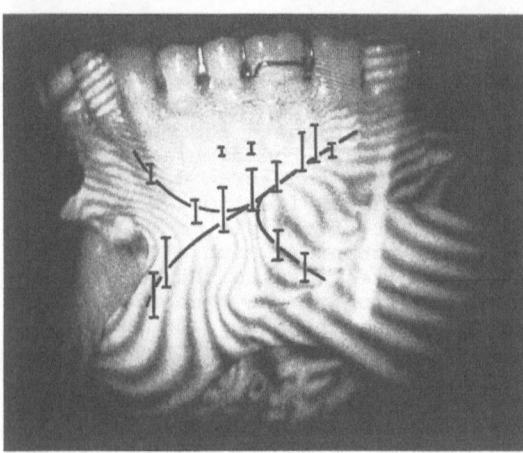

Fig. 2 Interferograms of the left and right maxilla recorded with pre-stress and expansion of 0.1 mm on each side (symmetrical plate)

Fig. 3 Lines of the stress distribution

inherent asymmetry of the anatomical specimen, and the bone reactions were therefore quantitatively unequal during the application of symmetrical lateral forces. The interferograms were analysed by measuring the density of the fringes along the line of a prominent diagonal and equidistant planes (Fig. 3). The gradient of the deformation was worked out from the pattern of relative deformation in each plane using graphical differentiation. By combining the points of local minima in the gradient of deformation in all the planes we obtained the lines along which stress is distributed.

The greatest movement was detected along the dental arch in the premolars, which, as a rule, showed a lateral displacement some 3-10 times greater than that of the frontal teeth. A movement 1-3 times smaller than that of the premolars was shown by the first and second molars and the canine teeth. An interesting movement was detected on the third molar, which was not in contact with the appliance; it always showed a movement in conformity with that of its iugum alveolare and a part of the infratemporal side of the maxilla above it (Fig. 4). The whole complex moved as a unit, the amount of the movement being propertional to the degree of aperture of the appliance. This means that the

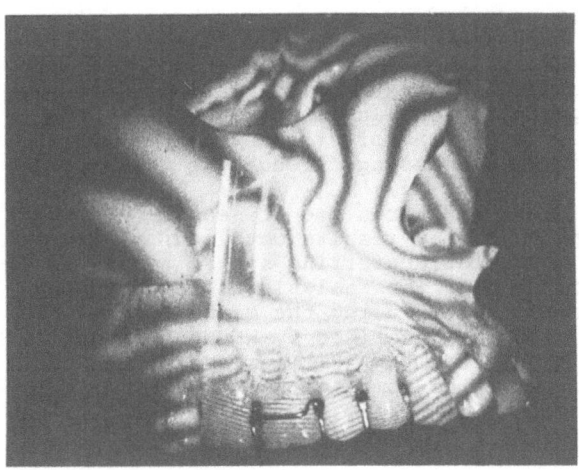

Fig. 4 Interferogram of the right maxilla without pre-stress,
expansion of 0.033 mm to each side (symmetrical plate)

movement was a passive one, caused by the movement of the dental
arch, the processus alveolaris and the apical base of the maxilla
in lateral direction.

The next greatest displacements, according to the gradient
of deformation of the processus alveolaris, could be detected
in the region from the first premolar to the second molar. This
region is the most dislocated part of the maxilla and it stret-
ches up to the apical base above the first and second premolars,
whereas the upper limit above the first and second molars is
much lower towards the margin of the alveolae. By tracing out
local minima of the gradient of deformation we obtained lines
which start below the level of the apical base, between the
first and second molars and the iugum alveolare of the canine
tooth and rises to the processus frontalis; and the left branch
rises to the processus zygomaticus. These results were found in
all the series of recorded interferograms. The best examples
could be detected in records with pre-stress.

Discussion

The response of the facial skeleton to the action of forces
applied during orthodontic treatment was analysed in our experiment under non-physiological conditions. Physiological masticatory
forces and their distribution along the upper jaw have been analysed using the methods described in /2,3,4,5,6/. Determination
of the deformations of such bones is possible using holographic
interferometry /1,7,8,9,10,12,13,14/. The lines of strain distribution during the masticatory function, determined by other
methods, were compared with those found in our holographic investigations. It could be shown that they are in general agreement. Reacting to the forces applied in our experiments the part
of the maxilla below the apical base moves in lateral direction,

and the resulting stress is absorbed by the compact bone of the processus zygomaticus and the processus frontalis. This leads to an inner static balance under new conditions. The main direction of stress transfer follows the two trajectories mentioned above. The only significant change during the action of lateral forces is the emergence of a transversal line of stress transfer, which joins both vertical trajectories, passing through the upper part of the fossa canina.

Teeth movements within the dental arch are of secondary importance in the interpretation of the results obtained, since their primary function is that of transferring stress to the bone. Apart from that our opinion is that the analysis of teeth movements alone can be carried out better using the method of holographic interferometry in vivo, as described by WEDENDAL and BJELKHAGEN /7,8,9,10/ and RYDEN et al. /11/. The results obtained in this way could possibly be combined with our results in further works.

Conclusions

The regions on the anterior plane of the maxilla, which are most strongly stressed by physiological mastication forces, are also affected the most by the laterally directed forces arising from the expansion of the jaw supplied with an active plate.

This analysis is, of course, not sufficient for explaining the complete inner strain distribution in the maxilla during the action of orthodontic expansions forces. However, in addition to the knowledge of reactions to masticatory forces, the results of our experiments contribute to the understanding of the intricate biomechanical principles of the dento-facial complex.

1. S. Vukicević, D. Vukicević, V. Nikolić, J. Hancević: Acta med. iug. 31, 251-259 (1977)
2. A. Benninghoff: Verh. Anat. Ges. 34, 189-205 (1925)
3. F. Pauwels: Z. Anat. Entwickl. Gesch. 114, 129-166 (1948)
4. B. Endo: J. Fac. Sc., Sec. V, III, 1-106, Tokyo (1966)
5. J. Keros: Mr. Sci. Theses, Zagreb (1978)
6. O. Muftić, J. Keros, K. Sivoncek: Proc. Cong. Anat. Assoc. Yug. (1977)
7. P.R. Wedendal, H.I. Bjelkhagen: Appl. Opt. 13, 2481-5 (1974)
8. P.R. Wedendal, H.I. Bjelkhagen: Acta Odont Scand 32, 189-199 (1974)
9. P.R. Wedendal, H.I. Bjelkhagen: Acta Odont Scand 32, 345-356 (1974)
10. P.R. Wedendal, H.I. Bjelkhagen: Acta Odont Scand 32, 131-145 (1974)
11. H. Ryden, H.I. Bjelkhagen, P.Ö. Söder: J. of Period. 46, 421-425 (1975)
12. P. Greguss: Opt. and Las. Tec. 8 (4), 153-159 (1976)
13. Z. Rajić: Dissertation, Zagreb (1976)
14. N. Abramson: Appl. Opt. 8, 1235-1240 (1969)
15. Steinbücher: Dissertation, München (1973)
16. T.M. Graber, B. Neuman: Removable Orthodontic Appliances, Saunders (1977)
17. V. Gazi-Coklica, V. Lapter: ASCRO, 10, 165-176 (1976)

IX. Holography in Otology

Holography in Otology

G. von Bally
Hals-Nasen-Ohrenklinik der Westfälischen Wilhelms-Universität
Kardinal-von-Galen-Ring 10, D-4400 Münster, Fed. Rep. of Germany

1. Introduction

Otology is the medical speciality, in which most holographic investigations were carried out so far [1,2,3].This may be caused by the fact that many of the structures related to the ear act as vibration transmitters. Thus, to study the function or dysfunction of these structures, the utility of methods for highly resolving, contactless, threedimensional vibration analysis like those of holographic interferometry is evident.

Since otological studies using holographic interferometry were realized mainly on the human ear, it seems to be appropriate to start with a brief description of its anatomy and physiology, as far as it is necessary for the explanation of these investigations.

The human peripheral hearing organ is embedded in the temporal bone, which includes the petrous pyramid forming a part of the skull base. A schematic cross section is shown in Fig.1.

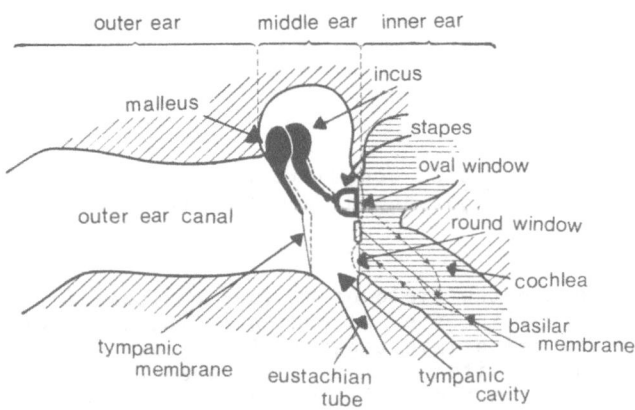

Fig.1. Schematic cross section of the human peripheral hearing organ.

183

The peripheral hearing organ can be divided into three parts:
the *outer ear* with the pinna (not included in the figure),
the outer ear canal, and the tympanic membrane, which forms the
boundary to
the *middle ear* being located within the tympanic cavity and containing the ossicular chain - malleus, part of which is connected
to the tympanic membrane (manubrium, refer to Fig.2), incus,
stapes -, as well as ligaments and muscles (not included in the
figure), and
the *inner ear* (cochlea) shaped like a coil. The cochlea is separated into two parallel canals by a membrane (basilar membrane),
the receptor cells for the hearing sense being arranged on it in
a special distribution.

Sound waves impinging through the outer ear canal on the tympanic membrane cause vibrations, which are transmitted by the
ossicular chain to the fluid system of the inner ear. The leverage
action of the ossicular chain results in the tympanic membrane
vibrations being transformed into movements of the stapes footplate of less volume displacement but greater force for impedance
matching purposes. The vibrations of the stapes footplate, attached to the oval window by an annular ligament, are propagating
through the lymphatic fluid in the cochlea, thus eliciting travel-
ling waves on the basilar membrane, which cause stimulation of the
receptor cells. Pressure equalization is achieved by the membrane
of the round window.

Figure 2 provides a closer look to the tympanic membrane. In
order to enable an easy orientation it is divided into four quadrants, as schematically outlined. Around the upper end of the
manubrium the tympanic membrane has a different structure and
lower tension (pars flaccida) than the main part of the membrane
(pars tensa).

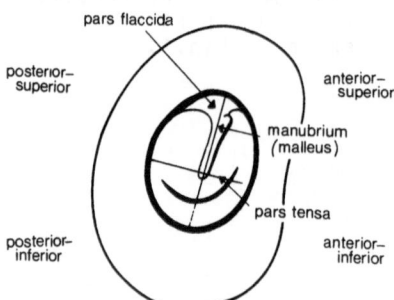

Fig.2   Schematic diagram of a
right human tympanic membrane.

Holographic applications in the field of otology reviewed in
the following paragraphs are classified according to the parts of
the sound conducting system being investigated. The sequence of
these structures chosen in opposite order to the propagation direction of the sound waves, from the skull and inner ear to the
tympanic membrane, shall provide an easy transition to the
following papers, dealing with holographic investigations on the
eardrum.

2. Holographic investigations on the skull and skull base

Vibration patterns of a macerated human skull, elicited by a bone conduction vibrator have been investigated using time-averaged holography [4]. These studies have been carried out to increase the knowledge of the mechanism of sound transmission by bone conduction.

Double-exposure holography has been used to investigate deformations of the human skull base under load [2]. The aim of these investigations has been to evaluate the stress distribution under load in the base of the skull, the knowledge of which may extend the understanding of the generation of typical micro-fractures of the petrous pyramids caused by accidents leading to hearing impairments. Figure 3 shows a reconstructed image from a corresponding double-exposure hologram. The skull was loaded in sagittal direction with a preload of 0.54 N and a small difference load between both laser illuminations of 0.02 N. The different deformation of the petrous pyramids in contrast to the surrounding parts of the skull base can clearly be recognized by the different number and direction of the corresponding holographic interference lines. Thus, in spite of problems as for instance the transferability of results of static investigations on isolated macerated bones to situations with dynamic load changes in living tissue and bones, holographic interferometry may be a good tool for basic investigations in accident research.

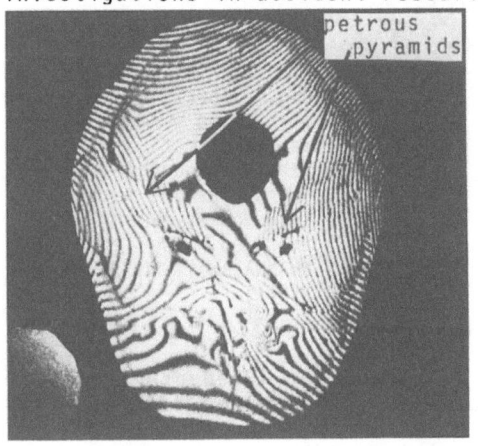

Fig.3 Deformation analysis of the human skull base subjected to sagittal load, using double-exposure holography, VON BALLY (1978).

3. Holographic investigations on the inner ear

When sound pressure levels (SPL) within the physiological range are applied to the ear, the movements of the basilar membrane are below the resolution limit of ordinary holographic interferometric techniques. For this reason and because of the fact that a surgical opening of the cochlea for a direct optical access to the basilar membrane affects the vibration behaviour of the inner ear structures seriously, enlarged scale models of the cochlea were used for holographic vibration analysis. In contrast to former holographic investigations on unrolled cochlear models [5], recent

studies were performed on a model with a coiled basilar membrane
[6]. In these experiments both travelling-wave and standing-wave
patterns were investigated and the influence of curvature on the
vibration modes of the basilar membrane was studied by using time-
averaged holography.

4. Holographic investigations on middle ear structures

By applying sound pressure via a specially fixed tube after removal of the outer and middle ear directly to the stapes footplate
in the oval window of anesthetized cats, time-averaged holograms
of the membrane of
the round window have
been recorded [7].Re-
constructed images
from corresponding
holograms are shown
in Fig.4, demonstra-
ting the variation of
the vibratory pattern
of the round window
membrane with
changing frequency.

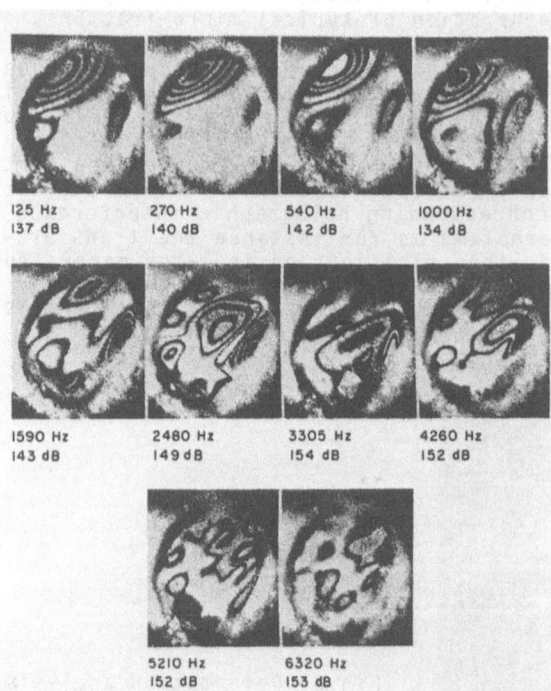

Fig.4 Vibration pattern
of a round window
membrane (cat) at dif-
ferent frequencies and
sound pressure levels,
recorded by time-
averaged holography,
KHANNA and TONNDORF
(1971).

From these holographic recordings inner ear impedance values
were calculated under the assumption of equal volume displacements
of the membranes of the oval and round window. The results con-
firmed those found earlier in capacitive probe measurements.

In order to investigate movements of the stapes footplate
time-averaged holography has been applied in experiments on fresh
human temporal bone preparations [8,9]. Thus, it was possible to
detect a small tilt movement of the stapes besides a piston like
oscillation, as demonstrated in Fig.5 by parallel interference
lines.

Contrast enhancement of the interference fringes was achieved
using a phase modulated reference wave [10]. This technique enables a selection of the fringe order in the holographic interferogram by phase adjustment.

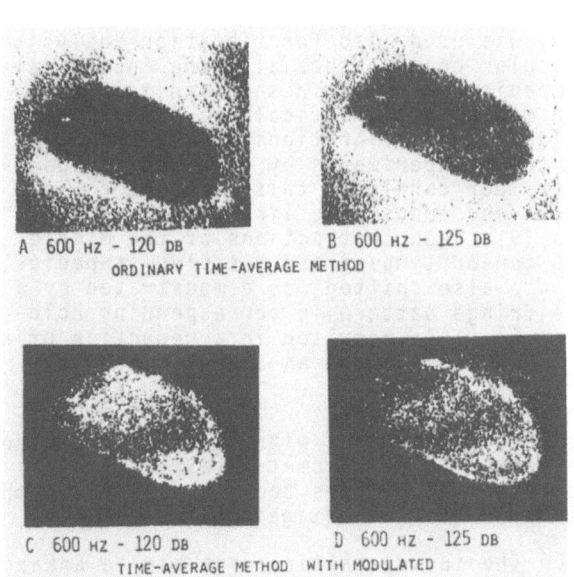

A 600 Hz - 120 DB   B 600 Hz - 125 DB
ORDINARY TIME-AVERAGE METHOD

C 600 Hz - 120 DB   D 600 Hz - 125 DB
TIME-AVERAGE METHOD WITH MODULATED
REFERENCE WAVE

<u>Fig.5</u> Movements of a human stapes footplate (inner ear side) investigated by different holographic techniques, HØGMOEN and GUNDERSEN (1976).

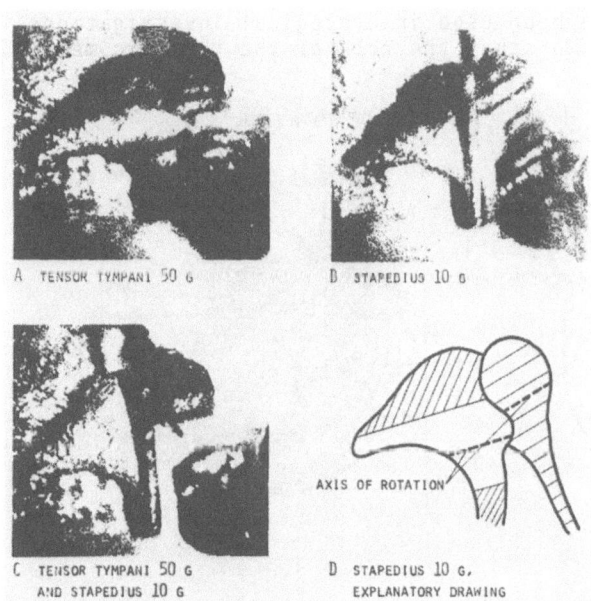

A TENSOR TYMPANI 50 G   B STAPEDIUS 10 G

C TENSOR TYMPANI 50 G   D STAPEDIUS 10 G,
AND STAPEDIUS 10 G   EXPLANATORY DRAWING

AXIS OF ROTATION

<u>Fig.6</u> Vibrations of human incus and malleus at 1 kHz and 124 dB SPL with forces applied to the middle ear muscles (to simulate corresponding contractions), studied by time-averaged holography, HØGMOEN and GUNDERSEN (1976).

Time-averaged holography was used also for vibration analysis
of other parts of the ossicular chain, especially the incudo-mallar joint. Opening the tympanic cavity of fresh human temporal
bone preparations from the medial side, optical access to incus
and malleus could be provided, while vibrations of the tympanic
membrane were elicited in these experiments by applying a closed
acoustic system to the outer ear canal. By this means it could
be demonstrated that malleus and incus move like a lever around
a frequency dependent axis [9]. When contractions of the middle
ear muscles, i.e. musculus tensor tympani and musculus stapedius,
were simulated, this axis was also shifted, as demonstrated by a
change of the interference fringe pattern in corresponding holographic interferograms (Fig.6). This resulted in a reduction of
the vibration amplitude, which is regarded as a protection mechanism for the inner ear.

Similar investigations were carried out with a different method
called electronic speckle pattern interferometry (ESPI) [9].
Using a TV-target as recording medium this technique provides the
possibility of holographic real-time investigations [11].

The described studies on the incudo-mallar joint aim to enhance
the knowledge of this part of the mechanical system of the peripheral hearing organ in order to develop new prostheses and techniques in tympanoplastic surgery.

5. Holographic investigations on the tympanic membrane

Holographic methods have been used in otological investigations
for the very first time to study the role of the tympanic mem-

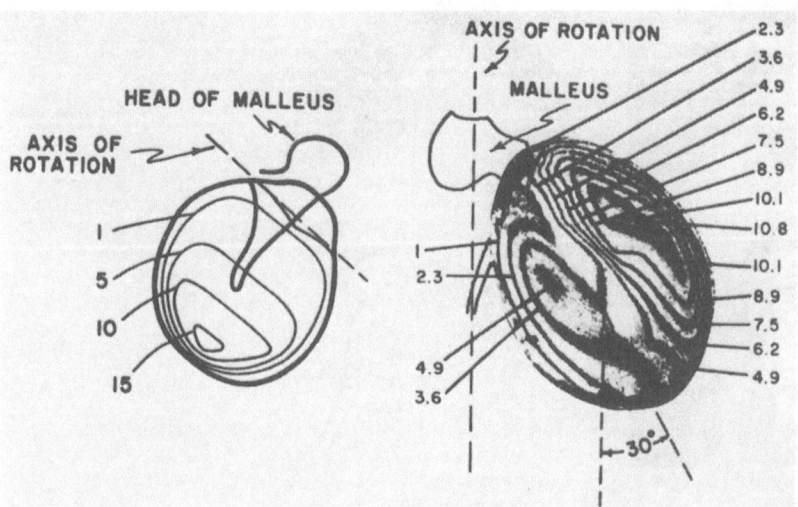

Fig.7 Vibration pattern of the tympanic membrane, left: capacitive probe
measurement in man by VON BEKESY, right: time-averaged holographic investigation in cat by KHANNA and TONNDORF (1972).

brane in middle ear transmission [12,13]. Starting with the application of time-averaged holography on tympanic membranes in cats, an outstanding controversy between two theories could be resolved [14]. Supported by capacitive probe measurements, the first one stated that the tympanic membrane vibrates like a hinged plate, as demonstrated by the drawing in the left part of Fig.7. In the second theory a more complicated vibration pattern was assumed (curved membrane concept), which could be revealed, using this holographic technique, as shown by the holographic interferogram on the right in Fig.7.

By means of time-averaged holography such complex vibration patterns of the tympanic membrane were found also in other animals like grasshoppers [15] and frogs [16].

Using the same holographic technique in studies on fresh human temporal bone preparations a vibration pattern of the tympanic membrane in man could be demonstrated, supporting the curved membrane concept instead of the hinged plate hypothesis [17]. From the volume displacement determined by these holographic investigations the impedance of the human tympanic membrane was calculated. A principle agreement with results of post-mortem experiments using an acoustic bridge was found.

Since time-averaged holography is not sensitive to the phase of the object vibration, the complex tympanic membrane impedance has to be calculated assuming a uniform phase distribution over the entire membrane. This must not be necessarily the case in biological membranes, especially at higher frequencies. Thus, it seemed to be interesting to look for suitable vibration-phase-sensitive methods, which are able to analyse for instance unsymmetric membrane vibrations.

Stroboscopic double-exposure holography turned out to be capable of investigating an unsymmetry of the oscillation of a membrane, damped on one side, in regard to its resting state [18], as demonstrated in Fig.8. The left interferogram in this figure shows the interference fringe pattern generated by superposition of the optical wave-fronts emanating from the membrane at its resting state and the maximum of its vibration towards the un-

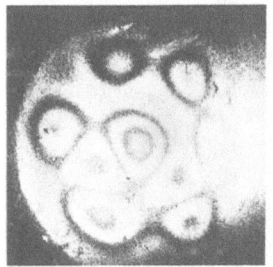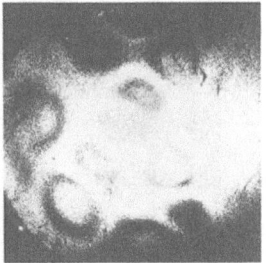

Fig.8 Holographic interference patterns on a membrane vibrating unsymmetrically to its resting state, left: resting state and undamped oscillation maximum exposed, right: resting state and damped oscillation maximum exposed, SIEGER, VON BALLY, RÖHLER (1978).

189

damped side. In the right interferogram the wavefronts according to resting state and maximum of the membrane vibration towards the damping material are brought to interference. The unsymmetry of the oscillation in respect to its resting state can clearly be recognized by the different numbers of interference fringes in each interferogram.

Using holographic subtraction an optical phase shift of π of the reference wave between the exposures of both selected vibration states results in a conversion of the intensity of the first order fringes, thereby increasing the visibility of these fringes by darkening of the background. Thus, a combination of the stroboscopic double-exposure technique and holographic subtraction enhances the detectability of displacement differences due to unsymmetric membrane vibrations at very small amplitudes, as demonstrated in experiments on a model of the human tympanic membrane [19]. Even investigations of arbitrary vibration waveforms and determination of the corresponding Fourier coefficients can be provided by the different techniques of temporally modulated holography [20].

Fast, non-periodic events, e.g. transient processes, can be studied by double-exposure holography using a Q-switched ruby laser. Thus, the vibrations of tympanic membranes of guinea pigs elicited by acoustic impulses were investigated in in-vitro experiments [21]. For that purpose a hologram recorded before and a second one taken at certain times after the start of the acoustic event were superimposed on the same holographic plate. Figure 9 shows the variation of the vibration pattern according

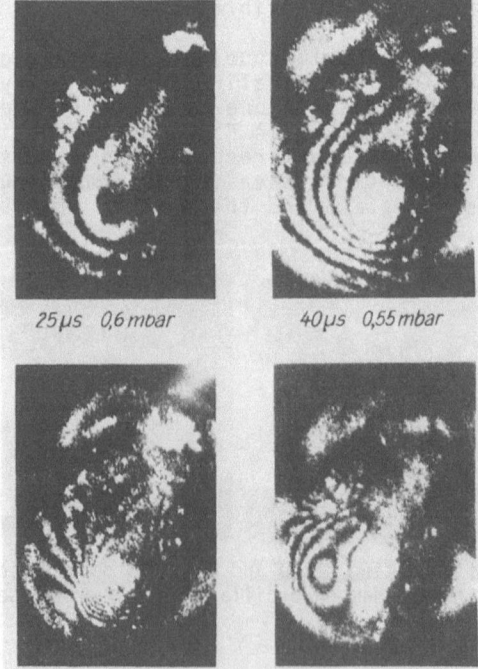

25 µs 0,6 mbar   40 µs 0,55 mbar

70 µs 0,5 mbar   100 µs 0,7 mbar

Fig.9 Vibration patterns of a tympanic membrane of a guinea pig subjected to acoustic impulses recorded by double-exposure holography using a pulsed ruby laser. The second laser pulse was released at the denoted times after the start of the indicated sound pressures, DANCER, FRANKE, SMIGIELSKI, FAGOT (1975).

to different delay-times of the second exposure after the beginning of the acoustic impulse. These investigations should
help to explain the occurrence of lesions after exposure to acoustic impulses mostly in a certain part of the tympanic membrane.

For steady-state vibration analysis by means of double-exposure
holography using a double-pulsed ruby laser system, both laser
pulses have to be released selectively in relation to the state
of oscillation. Advantages of this technique, that have caused
interest in otological applications, are applicability to in-vivo
investigations together with vibration-phase-sensitivity [2].
Due to the short pulse width and the short interval between both
pulses, problems arising from instability of the optical set-up
or the object are of minor importance in this case compared
to the use of gas lasers. Thus, the requirements of interferometric stability can be met even in the presence of parasitic
micromovements of living objects.

This technique has been used to investigate its capability in
studying the influence of the tympanic membrane vibrations on
the sound transfer function of the middle ear, as well as its
principle applicability to the control of healing processes in
tympanoplastic surgery, and the possibility of locating disorders
in the middle ear without removing the tympanic membrane. For
the latter purpose the basic idea was, that, since pathological
changes of the mechanical properties of the middle ear have an
influence on the vibratory pattern of the tympanic membrane, a
vibration analysis may provide the possibility of a differential
diagnosis of these dysfunctions without opening the tympanic cavity. This could be demonstrated in investigations on human temporal bone preparations [22,23,24].

In absence of macroscopically visible pathological alterations
a typical vibration pattern of the human tympanic membrane, cha-
racteristicly depending on frequency and phase of the eliciting
sound pressure oscillation could be revealed. Figure 10 shows the
variation of the vibration mode of a left tympanic membrane according to changes in frequency at constant sound pressure level
and the laser pulses being released at a constant phase relation
to the sound pressure oscillation. The vibration amplitudes of the
manubrium, here positioned from the upper left corner to the
center of each interferogram, are considerably smaller than those
of adjacent parts of the tympanic membrane, including that of
pars ˙flaccida around the upper end of the manubrium. Up to a frequency of 2 kHz a larger displacement in both posterior quadrants
compared to the movement in the anterior ones can be detected. At
higher frequencies the vibratory pattern becomes more intricate
and is separated into several additional sections. These findings
correspond to studies using different holographic techniques
[17,25].

Since double-exposure holography is sensitive to the phase of
the object vibration, this technique provides the possibility to
investigate the phase relation between sound pressure oscillation
and corresponding tympanic membrane vibration. The variation of
the vibration pattern due to changes of the releasing phases of
the laser pulses in relation to the sound pressure oscillation,

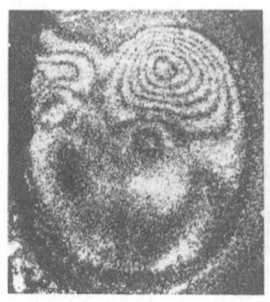 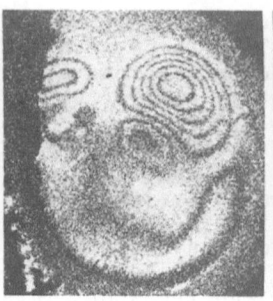 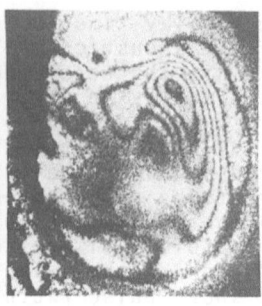

1.5 kHz   2.0 kHz   3.0 kHz

Fig.10 Frequency dependent variation of the vibration pattern
of a left human tympanic membrane recorded by double-exposure
holography at a constant sound pressure level of 105 dB and release of the laser pulses at consecutive extreme values of the
sound pressure oscillation (90°/270°), VON BALLY (1976).

is demonstrated in Fig.11 for two frequencies. As indicated by
the symbols in the central column both laser pulses are released
at an interval of half a period, while being shifted in steps of
45° relative to the sinusoidal sound'pressure oscillation in the
plane of the tympanic membrane for each double-exposure record.
Holographic interferograms of the same tympanic membrane as shown
in Fig.10, demonstrate at a frequency of 2 kHz a variation of tympanic membrane displacements, characterized by the different
numbers of interference fringes (left column in Fig.11), in accordance with the changes of sound pressure differences corresponding to the phases of the sound pressure oscillation, at which
the laser pulses are released (central column in Fig.11). Assuming
a sinusoidal membrane vibration, a coincidence in phase of tympanic membrane vibration and sound pressure oscillation can be
gathered from these investigations [23].

At a frequency of 3 kHz such a conformity in phase of the oscillations of the tympanic membrane and the eliciting sound pressure
is lost (right column in Fig.11). A difference in phase can be deduced e.g. from the fact that, even when both laser pulses are released at successive zero passages of the sound pressure oscillation, i.e. no difference in sound pressure at the releasing
phases of both laser pulses, a considerable number of interference
fringes can be detected.

Characteristic deviations from the described normal vibration
pattern of the human tympanic membrane could be detected caused
by alterations of the mechanical properties of the middle ear.

An example for the influence of artificially introduced modifications of the middle ear mechanics, is presented in Fig.12. Here
a comparison of the vibration patterns of a left human tympanic
membrane demonstrates that after resection of the incus - contrary

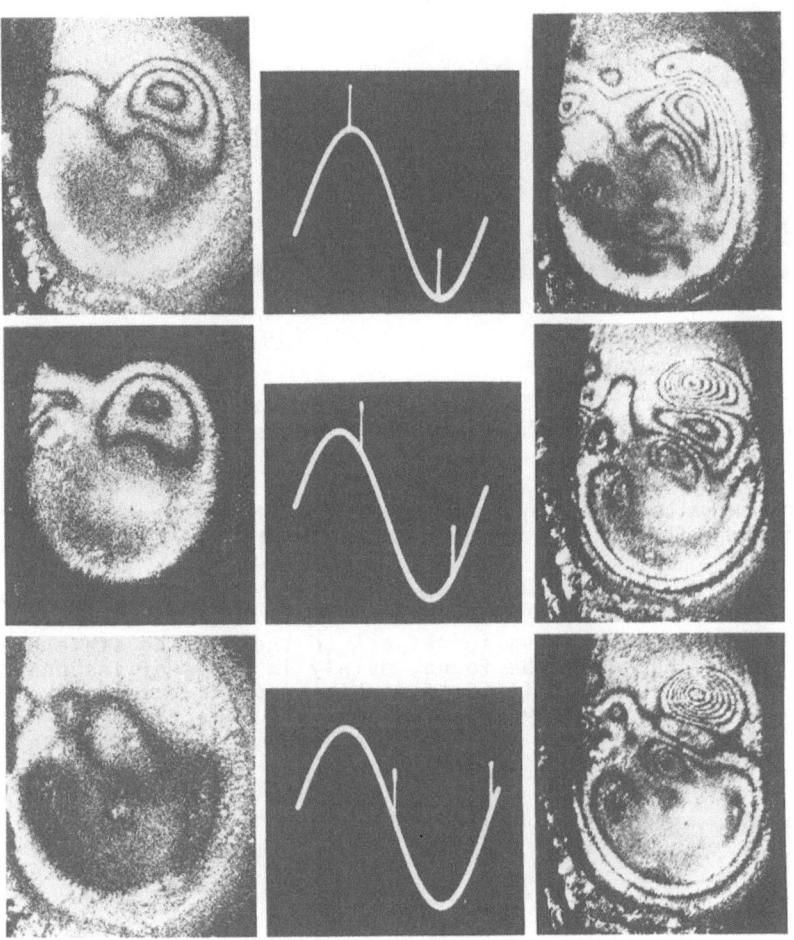

2.0 kHz
100 dB

3.0 kHz
105 dB

Fig.11 Comparison of changes of the vibration pattern of a left human tympanic membrane for two frequencies due to variation of the phases of the eliciting sound pressure oscillation, at which the laser pulses are released. VON BALLY (1977).

to the normal vibratory pattern - the oscillation amplitude of the anteerior quadrants equals that of the posterior ones.

Pus in the middle ear caused by a serous otitis media results in an intricate vibration pattern of the tympanic membrane with several small, separately vibrating sections, even at frequencies below 2.0 kHz, as shown in Fig.13.

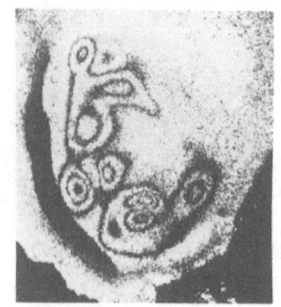

Fig.12   Fig.13

Fig.12 Vibration pattern of a left human tympanic membrane (2.0 kHz; 110 dB; 270°/450°) (left) before and (right) after removal of the incus, VON BALLY (1976).

Fig.13 Vibration pattern of a right human tympanic membrane affected by a serous otitis media (1.5 kHz; 105 dB; 270°/450°), VON BALLY (1976).

Unsymmetry of the vibration in respect to the resting state of the tympanic membrane could be found, mainly in cases of lesions within pars tensa. The capability of double-exposure holography to investigate such vibratory unsymmetries is demonstrated in Fig.14. While the difference in eliciting sound pressure between

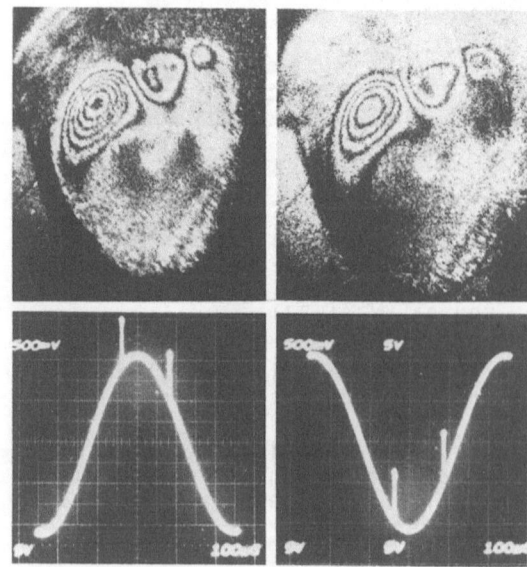

Fig.14 Vibration pattern of a right human tympanic membrane perforated at its inferior rim (1.0 kHz; 115 dB), analysis of an unsymmetrical oscillation of the membrane using double-exposure holography; VON BALLY (1977).

both illuminated vibration states of the tympanic membrane is kept constant for each interferogram, the releasing phases of the laser pulses are shifted by 180° in relation to the sound pressure oscillation, as can be gathered from the oscilloscope records in Fig.14. The resulting interference patterns, however, indicate a significant displacement difference by the different numbers of interference fringes in corresponding particular sections of the membrane vibration.

Nodal lines indicating a phase-opposition of the vibration of adjacent areas even at frequencies below 2 kHz could be detected in several cases of pathological alterations like perforated tympanic membranes and disconnected incudo-mallar joints [23]. An example is presented in Fig.15, showing bright nodal lines seaparating three sections of a left tympanic membrane vibrating in phase-opposition.

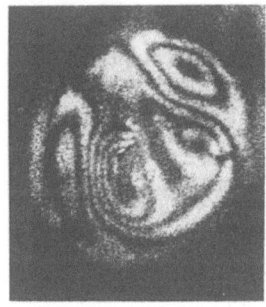

Fig.15 Vibration pattern of a left human tympanic membrane after disconnection of the incudo-mallar joint by resecting the head of the malleus, indicating a phase opposition by nodal lines separating three vibratory sections of the membrane (1.0 kHz; 105 dB; 270°/450°) VON BALLY (1977).

The interest in such investigations arises from the fact, that calculations of the complex middle ear impedance from the volume displacement of the tympanic membrane so far are based on the assumption of a uniform vibration phase distribution on the entire membrane, especially at low frequencies [17].

6. Conclusions

First holographic investigations in otology were carried out on animals using the time-averaged method. Since then a variety of other holographic techniques has been applied, too, including modern methods like temporally modulated holography. The main interest in otological applications of holography has turned from studies on animals to clinical use in man. Thus, by using holographic techniques new results could be obtained in basic research on the sound transfer function of different structures of the human peripheral hearing organ as well as the principle applicability of certain of these methods to otological diagnostics be demonstrated experimentally.

As far as investigations in man are concerned holographic studies, thus far described in this review, were performed on human temporal bone preparations surgically treated in order to provide an optical access to the structures under consideration. Subsequent new developments to holographic in-vivo investigations are reported on in the following three papers of this issue.

Summarizing this survey otology seems to be a good example for demonstrating that holographic methods can be used in certain medical fields already to a considerable extent.

7. References

1   G. von Bally: Holographic methods in biomedical sciences, Pre-prints Int. Conf. on Opt. Comput. in Res. and Develop., Visegrad, 2, 68 (1977).
2   G. von Bally: Holografische Verfahren in Medizin und Biologie, Frühjahrsschule "Holografische Interferometrie in Technik und Medizin", Hannover, (1978).
3   M. Hoke, G. von Bally (eds.): Proc. Symp. 1976 Spec. Res. Area 88 and Conf. on Electrocochleography and Holography in Medicine, Münster, (1976).
4   Y. Ogura et al.: A holographic study of the human skull vibration, Audiol. Jap. 19, 163 (1976).
5   R. Chadwick, J.D. Cole: Hydroelastic modes in the cochlea, IUTAM Conf., Delft, (1976).
6   G. Edlund. G. Wikander, N.-E. Molin: The visualization of modes in a circular cochlear model by hologram interfero-metry, J. of Sound and Vibration 59, 299 (1978).
7   S.M. Khanna, J. Tonndorf: The vibratory pattern of the round window in cats. J. Acoust. Soc. Amer. 50, 1475 (1971).
8   Y. Ogura, Y. Masuda, M. Miki, T. Takeda, S. Watanabe, T. Ogawara, S. Shibata, T. Uyemura, Y. Yamamoto: Vibration analysis of the human skull and auditory ossicles by holo-graphic interferometry, this issue.
9   K. Høgmoen, T. Gundersen: Holographic vibration analysis of the ossicular chain, see [3], 1, 247 (1976).
10 K. Høgmoen, T. Gundersen: Holographic investigation of stapes footplate movements, Acustica 37, 198 (1977).
11 O.J. Løkberg, K. Høgmoen, T. Gundersen: Use of ESPI to measure the vibration of the human eardrum in-vivo and other bio-logical movements, this issue.
12 S.M. Khanna: A holographic study of tympanic membrane vibra-tions in cats, Ph.D.-Thesis, City University of New York, (1970).
13 J. Tonndorf, S.M. Khanna: The role of the tympanic membrane in middle ear transmission, Ann. Otorhinolaryngol. 79, 743 (1970).
14 S.M. Khanna, J. Tonndorf: Tympanic membrane vibrations in cats studied by time-averaged holography, J. Acoust. Soc. Amer. 51, 1904 (1972).
15 A. Michelsen: The physiology of the Locust ear, II. frequency discrimination based upon resonances in the tympanum, Z. vergl. Physiol. 71, 63 (1971).
16 J. Mazumdar, D. Bucco, C. Hansen: Time-averaged holography for the study of the vibrations of the tympanic membrane in frog cadaver ears studied by time-averaged holography, J. Acoust. Soc. Amer. 52, 1221 (1972).
17 J. Tonndorf, S.M. Khanna: Tympanic-membrane vibrations in human cadaver ears studies by time-averaged holography, J. Acoust. Soc. Amer. 52, 1221 (1972).

18 R. Röhler, C. Sieger: Analysis of unsymmetrical membrane vibrations by holographic interferometry, Opt. Commun. <u>25</u>, 297 (1978).
19 C. Sieger, G. von Bally, R. Röhler: Holografische Interferometrie zur Schwingungsanalyse des Trommelfelles, Frühjahrsschule "Holografische Interferometrie in Technik und Medizin", Hannover, (1978).
20 C. Sieger, R. Röhler: Measurement of vibration waveforms using temporally modulated holography, this issue.
21 A.L. Dancer, R.B. Franke, P. Smigielski, F. Fagot: Holographic interferometry applied to the investigation of tympanic-membrane displacements in guinea pig ears subjected to acoustic impulses, J. Acoust. Soc. Amer. <u>58</u>, 223 (1975).
22 G. von Bally: Untersuchungen der Schwingungen menschlicher Trommelfellpräparate mit Hilfe von Doppelpulsholografie, see [3], 1, 263 (1976).
23 G. von Bally: Holographic analysis of tympanic membrane vibrations in human temporal bone preparations using a double pulsed ruby laser system, in : E. Marom, A.A. Friesem, E. Wiener (eds.): Proc. Int. Conf. Appl. Hol. and Opt. Data Process, Pergamon Press, 593 (1977).
24 G. von Bally: Holografische Schwingungsanalyse des Trommelfelles, Laryng. Rhinol. <u>57</u>, 444 (1978).
25 W. Fritze, K. Burian, O. Schwomma: Holografische Untersuchungen des Trommelfelles, see [3], <u>1</u>, 239 (1976).

Otological Investigations in Living Man Using Holographic Interferometry

G. von Bally

Hals-Nasen-Ohrenklinik der Westfälischen Wilhelms-Universität
Kardinal-von-Galen-Ring 10, D-4400 Münster, Fed. Rep. of Germany

1. Introduction

After revealing the utility of double-exposure holography using a pulsed ruby laser system for studies on the vibratory pattern of tympanic membranes in human temporal bone preparations [1,2], the applicability of this technique in living man was investigated [3,4]. One of the objectives of these investigations was the study of the transferability of results obtained in in-vitro experiments to the function of the tympanic membrane in-vivo. In order to evaluate the clinical applicability of this holographic technique, besides investigating tympanic membrane vibrations in cases of a regular middle ear status, first studies on the influence of pathological alterations within the middle ear on the vibratory pattern of the eardrum should be carried out in living man.

2. Technical arrangements

In-vivo studies on the human tympanic membrane using double-exposure holography are complicated by the difficult optical access through the narrow and somewhat curved outer ear canal and by the requirement to determine amplitude and phase of the eliciting sound pressure oscillation in the plane of the eardrum for a correct interpretation of the resulting interference fringe pattern. For the purpose of recording double-exposure interferograms of the tympanic membrane in man, a special closed acoustic system was developed, which allows measurement of the phase of the eliciting sound pressure oscillation at the location of the tympanic membrane. The principal function and a prototype of the apparatus are shown in Fig.1. A beam splitting mirror separates the parts of the object beam impinging onto and reflected by the tympanic membrane, which have the same optical axis but opposite direction. Tympanic membrane vibrations are elicited by a loudspeaker (earphone) connected to the closed acoustic system. A probe microphone monitors the sound pressure oscillation and is used to trigger the laser pulses. It is calibrated to measure the sound pressure level and phase in the plane of the tympanic membrane. A usual speculum acts as an adapter between the closed acoustic system and the outer ear canal. For phase investigations of the sound pressure oscillation it was proved that this system has no resonances at the frequencies used.

The principle of the electronic remote control for phase selective laser pulse release in relation to the sound pressure oscil-

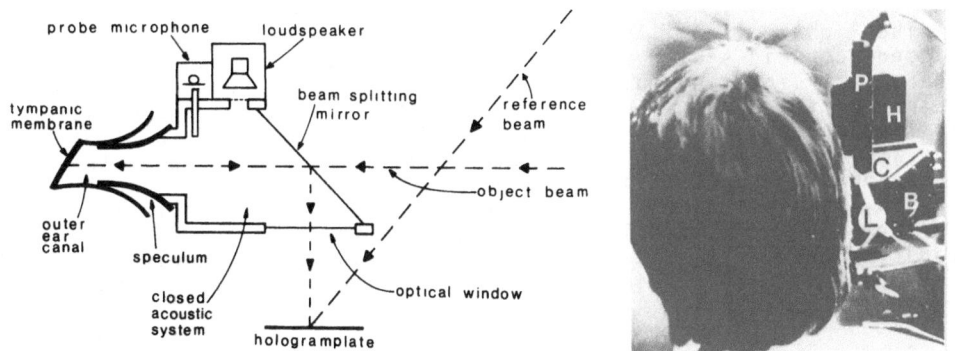

Fig.1 Closed acoustic system for sound application and holographic recording of tympanic membrane vibrations through the outer ear canal; left: scheme of the system; right: prototype of the apparatus applied to a patient, P - probe microphone, H - hologram plate holder, C - closed acoustic system, B - beam splitting mirror, L - loudspeaker (earphone).

lation is outlined in Fig.2. Sound pressure vibrations generated by an earphone attached to the closed acoustic system, are picked up by a probe microphone, the preamplified output signal of which is fed to a sound level meter as well as to an input of a storage oscilloscope. After passing through a limiting amplifier this signal can be variably phase delayed, and is then used for phase controlled triggering of a so called double-pulse electronic device releasing the laser pulses.

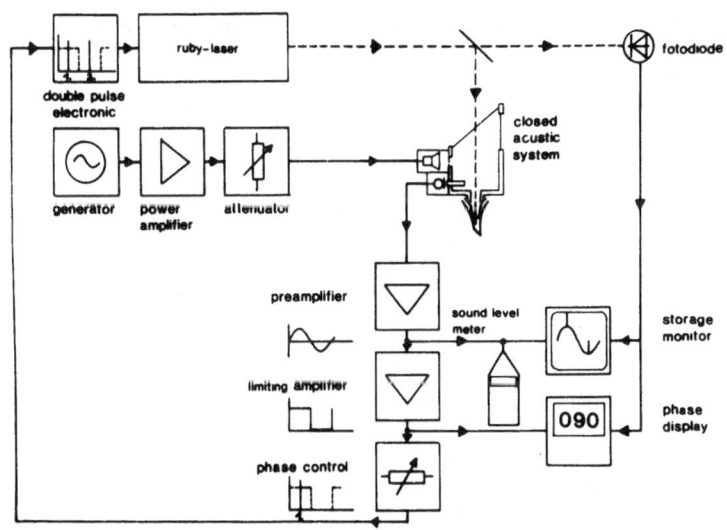

Fig.2 Scheme of the electronic control for phase selective laser pulse release in relation to the sound pressure oscillation.

199

The light pulses illuminate the tympanic membrane via the
closed acoustic system as well as a photodiode simultaneously.
The output signal of this photodiode is added to the preamplified
output of the probe microphone and monitored on a storage oscilloscope screen, as well as compared with the output of the limiting amplifier, in order to calculate the releasing phase of the
laser pulses in relation to the eliciting sound pressure oscillation, thus enabling a direct check of the phase control setting.
Correct sound pressure level and phase in the plane of the tympanic membrane are adjusted after multiplying sound level meter,
and phase display readings, respectively, with frequency depending
calibration factors.

During each holographic recording procedure sound is applied
for about 15 s. Before starting a series of records, the laser
output energy density is measured in the plane of the tympanic
membrane, and it is checked that it does not exceed $1mJ/cm^2$,
which can be regarded as far enough below a corresponding safety
level. No parasatic oscillations of the rigid optical set-up
could be detected in the recorded holographic interferograms within the range of the applied sound pressure levels.

3. Otological investigations in-vivo

Figure 3 shows holographic interferograms of the vibration pattern
of a right tympanic membrane of a young man with regular middle
ear status at a sound pressure level of 100 dB and different frequencies, recorded with the described equipment. Both laser pulses
are released at the peak values of the sound pressure oscillation.
It can be recognized that above 1.5 kHz - similar to preparation
experiments [1,5] - the intererence fringe pattern becomes more
irregular, i.e. the posterior quadrants are not any longer vibrating uniformly.

Double-exposure interferograms from the same tympanic membrane
as in Fig.3 are presented in Fig.4 demonstrating the variation

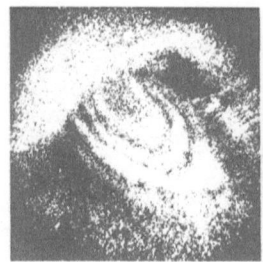 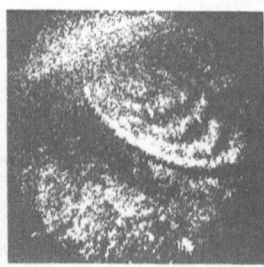 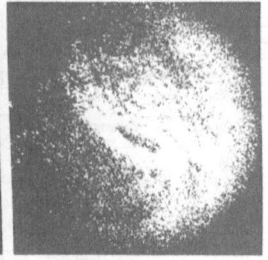

1.0 kHz   1.5 kHz   3.0 kHz
100 dB; 270°/450°

Fig.3 Vibration pattern of a right tympanic membrane of a young
man with regular middle ear status at constant sound pressure
level and releasing of the laser pulses at successive extreme
values of the sound pressure oscillation, but at different frequencies, as indicated.

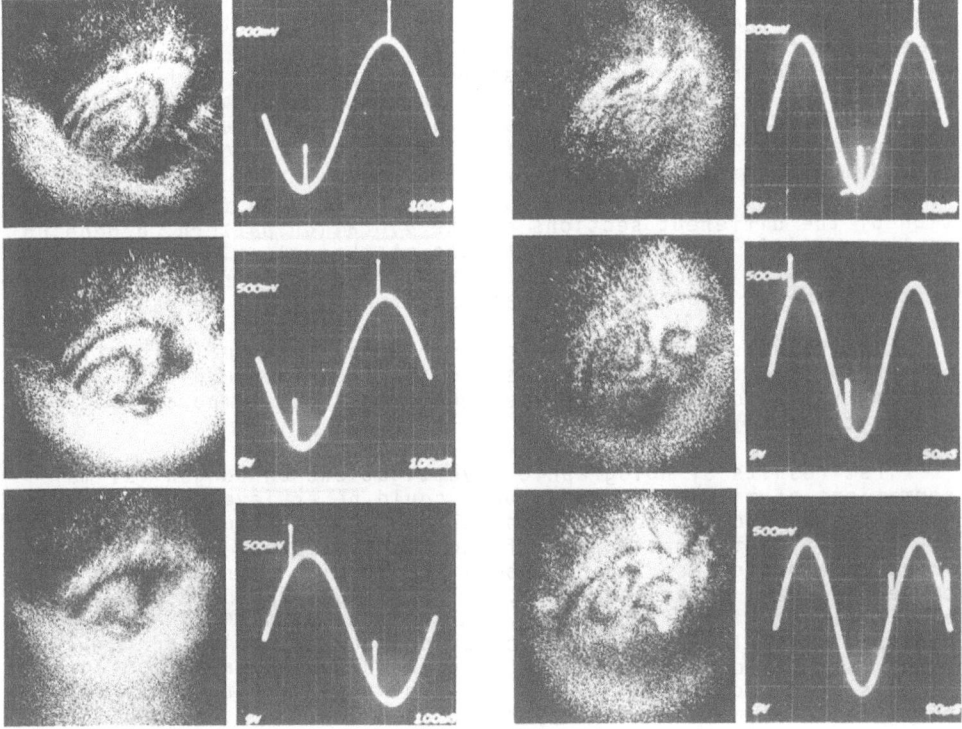

1.0 kHz   3.0 kHz

Fig.4 Variation of the vibration pattern of a right tympanic membrane (same subject as in Fig. 3) for two frequencies, at constant sound pressure level of 100 dB, but the laser pulses being released at different phases of the sinusoidal sound pressure oscillation.

of the vibratory pattern at two frequencies, but here recorded with the laser pulses being released at different phases of the sinusoidal sound pressure oscillation, as can be gathered from the photographs of the oscilloscope screen. The sound pressure level is kept constant to 100 dB. In conformity with cadaver experiments, at a frequency of 1 kHz most interference fringes, i.e. maximal membrane displacement, can be found when both laser pulses are released at successive peak points of the sound pressure (left column in Fig.4). The displacement is then reduced according to the actual releasing phase variation, which correspond to changes of sound pressure differences between both illuminated states of the tympanic membrane vibrations. Thus, it can be deduced that within the range of significance of this method at 1 kHz membrane and sound pressure are vibrating in phase.

Such a coincidence in phase of the vibration of the tympanic membrane and the sound pressure can no longer be detected at a frequency of 3 kHz (right column in Fig.4). In this case even when no difference in sound pressure exists between both illuminated states of the membrane vibration, e.g. both laser pulses being released in the zero passages of the sound pressure oscillation, a considerable number of interference fringes are generated, demonstrating that sound pressure and membrane vibration are out of phase at 3 kHz. Moreover, it can be deduced from the variation of the different sections of the vibration pattern, depending on the specific sound pressure phases at the releasing points of the laser pulses, that at this frequency even different partitions of the tympanic membrane are phase shifted relative to each other.

These results correspond in principle to the normal vibration pattern of the tympanic membrane and its frequency and phase depending variations found in experiments on human temporal bone preparations [1].

First results of holographic investigations on congenital malformations of middle ear structures could be achieved in patients using the method and equipment described above. Figure 5 shows in its left part a photograph of the right tympanic membrane of a young male patient suffering from a thalidomide embryopathy recorded by means of an endoscope. Malformations can be recognized, e.g. a dislocation of the manubrium in anterior direction and an abnormal chord-like structure posterior and perpendicular to the manubrium. The consistence of such structures can vary from an osseous over a cartilaginous to a ligamentous texture [6]. Its position in relation to the tympanic membrane can be very different, too. It may run a short distance behind, or be adhesively connected to the membrane. Sometimes findings in these respects are ambiguous, when using a speculum or a SIEGLE pneumatic otoscope (an otological instrument, that by closing the outer ear canal air-tightly allows inspection of tympanic membrane movements due to artificial air pressure changes). Also an endoscopic photograph cannot always give evidence to such questions.

The vibration pattern of this tympanic membrane was investigated holographically. A holographic interferogram, recorded at 100 dB sound pressure level, a frequency of 2 kHz, and the laser pulses being released at the peak points of the eliciting sound pressure oscillation, is presented for comparison in the right section of Fig.5, besides the endoscopic photograph. The reflec-

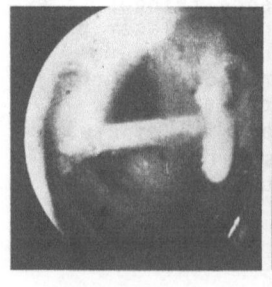 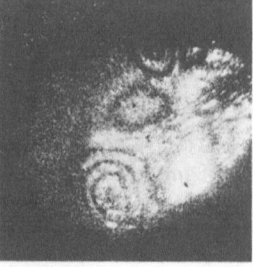

Fig.5 Right tympanic membrane of a young male patient with malformed middle ear structures due to thalidomide embryopathy, left: endoscopic photograph, right: holographic interferogram (2 kHz; 100 dB; 90°/270°).

tion of the manubrium as well as that of the perpendicularly positioned structure can clearly be recognized in the reconstructed
image. This structure separates the normally at 2 kHz uniformly
vibrating posterior quadrants into two separately oscillating
sections. The chord-like structure itself as well as the manubrium
and the surrounding parts of the tympanic membrane appear bright
and not interrupted by dark interference lines up to the rim of
pars tensa, which is fixed and does not vibrate. Thus, it can be
concluded, that this structure is in contact with the membrane
and does therefore not allow a detectable movement of that part
of the tympanic membrane at 100 dB sound pressure level, within
the range of significance of this method. Another partition in
the vibration of the tympanic membrane can be detected around the
upper end of the manubrium. As known from cadaver experiments
these interference fringes characterize the vibration of pars
flaccida. In the holographic interferogram this oscillation can
clearly be separated from the vibrating section in the posterior-
superior part of pars tensa. Such a distinction cannot be ascer-
tained in the endoscopic photograph.

The variation of the interference pattern of this tympanic membrane according to changes of sound pressure level at constant
frequency and release of the laser pulses at consecutive extreme
values of the sound pressure oscillation are demonstrated in
Fig.6. The vibration mode described above remains unchanged up to a
sound pressure level of 105 dB. The number of interference fringes
in the area of pars flaccida can be recognized as being higher
than that in the posterior-superior area of pars tensa. At 110 dB
a vibration of the whole posterior part of the membrane can be
detected by an interference fringe enclosing all other sections
of the vibration pattern. The left boundary of this fringe is in
conformity with the bright area in the upper-left part of the endoscopic photograph in Fig.5, thus demonstrating that this structure is connected to the tympanic membrane and does not vibrate,
even at this high sound pressure level, which cannot be deduced
from the endoscopic picture. The separation of the posterior inferior and superior vibration of pars tensa as well as that of
pars flaccida is still visible.

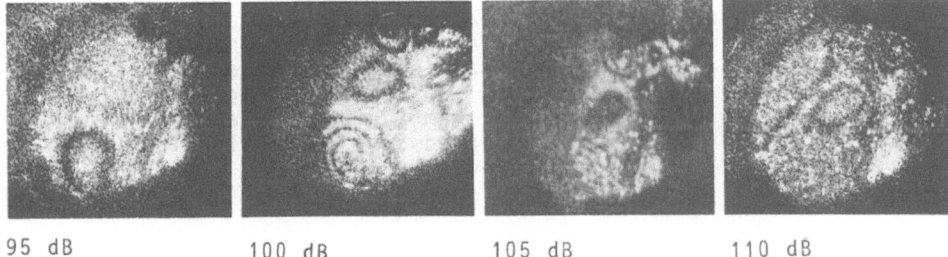

95 dB   100 dB   105 dB   110 dB

Fig.6 Vibration pattern of the same tympanic membrane as shown
in Fig.5, but a different sound pressure levels (2 kHz; 90°/270°).

4. Further technical developments of the equipment

For clinical routine applications of the described technique a quick generation of the holograms without a wet development of a photographic plate as well as a possibility of positioning a flexible optical system in relation to the patients, contrary to the actual use of a rigid set-up, would be of advantage. For that purpose a closed acoustic system with a light guide using a standard endoscope fiber bundle for the illuminating object beam was developed (left picture in Fig.7) and holograms were recorded on a thermoplastic film material, which enables dry and in-situ development within about 10 s. An example is shown in the right section of Fig.7, demonstrating the vibration pattern of an artificial membrane - a prototype of a material for temporal substitution of tympanic membrane tissue in tympanoplastic surgery [7] - within a plane scale model of the human outer ear canal. A sufficient picture quality can be achieved although the double-exposure holograms are generated by guiding the output pulses of a ruby laser via the standard endoscope fiber bundle and recording on the thermoplastic film material. Thus, such holograms can be obtained within 10 s and may then be displayed on a video monitor and stored on a videotape, or picked up by a polaroid camera for documentation purposes.

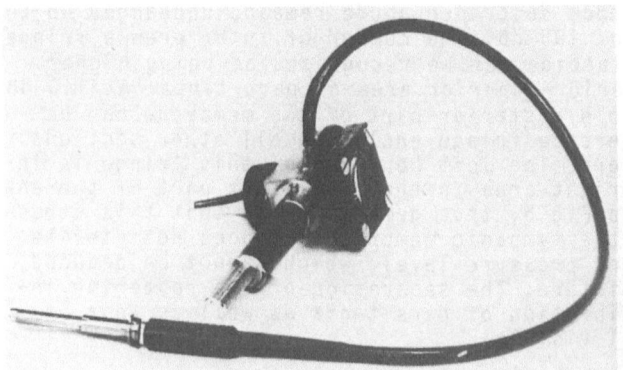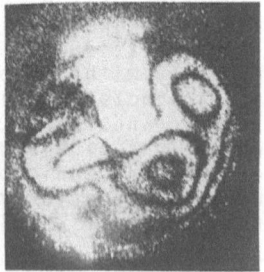

Fig.7 Left: closed acoustic system for sound application and holographic recording of tympanic membrane vibrations through the outer ear canal, using a standard endoscope fiber bundle as a light guide for the illuminating object beam; right: double-exposure hologram of an artificial membrane in a plane scale model of the human outer ear canal, recorded on a thermoplastic film by means of this system using a pulsed ruby laser.

5. References

1 G. von Bally, in: E. Marom, A.A. Friesem, E. Wiener (eds.):
  Appl. of Hol. and Opt. Data Process., Pergamon Press,
  593 (1977)
2 G. von Bally, elsewhere in this issue
3 G. von Bally, in: Frühjahrsschule 78, "Holographische Interferometrie in Technik und Medizin", Hannover, (1978)
4 C. Sieger, G. von Bally, R.Röhler, in: Frühjahrsschule 78,
  "Holographische Interferometrie in Technik und Medizin",
  Hannover,(1978)
5 J. Tonndorf, S.M. Khanna, J. Acoust. Soc. Amer. $\underline{52}$, 1221 (1972)
6 G. von Bally, S. Baumeister, Arch. Oto-Rhino-Laryng. (1979)
  (in press)
7 In cooperation with L. Feenstra, University of Amsterdam,
  Dept. of Otorhinolaryngology, J. Feijn, F. Kohn, Twente
  University of Technology, Dept. of Chemical Engineering,
  Enschede, The Netherlands

This work was supported by grants of the Deutsche Forschungsgemeinschaft.

On Holographic-Interferometric Investigations of the Membrana Tympani (Living Man)

W. Fritze
2nd Dept. for ENT, University of Vienna, Austria

H. Kreitlow and D. Winter
Institut für Meßtechnik im Maschinenbau, Technische Universität
D-3000 Hannover, Fed. Rep. of Germany

The form and amplitude of the vibration of the human eardrum until now has been primarily investigated on the isolated human temporal bone and on living experimental animals. The contemporary approved theory that the umbo has the same amplitude as the surrounding parts of the eardrum originates from v. BÉKÉSY (1941); Fig.1. This is hardly understandable technically because the umbo represents the main mechanical impedance. Its amplitude therefore should be reduced as compared to the surrounding eardrum. The expected form of vibration should be similar to that of an eardrum with an acute otitis media (with secretion-stowing in the middle-ear); in this case the maximal excursion is in the largest free area, i.e. in the posterior portion. Corresponding modes of vibration have been demonstrated by several holographers on the isolated temporal bone (TONNDORF and KHANNA, 1972; FRITZE, BURIAN and SCHWOMMA, 1976; v.BALLY, 1977); Fig. 2. Investigation in vivo is necessary, however, because the preparation of the isolated specimen has shown to influence the hologram. Moisture is an especially important factor: dryness on the one hand and absorption of irrigation-fluid when fraising on the other hand which can cause distention of the specimen.

The amplitude of the vibration of the eardrum in the upper physiological range is in the same magnitude as that of the wavelength of visible light. The holographic interferometry offers a suitable procedure because the modes of vibration can be observed simultaneously over the whole area. Whereas a point-by-point examination (for example a capacity-probe; v. BÉKÉSY, 1941) does not give such a satisfactory overall-view. A real-time technique (FRITZE, BURIAN and SCHWOMMA,1976)

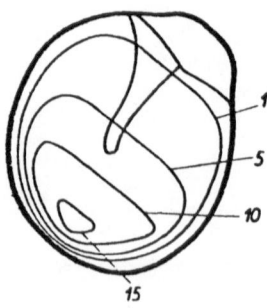

Fig.1 Relative amplitudes of the vibration, valid for frequencies up to 2 kcps. Temporal bone preparation, right side. From v. BÉKÉSY, 1941.

gives very good information but it is hardly applicable in vivo because of micro-movements of the living body. When using a double-pulsed laser, the relative relaxed position does not have to be longer than the interval of the pulses; this is usually half a wavelength of the applied frequency.

A ruby-laser with a capacity of 1o mJ was used (fig.3). This represents an absolutely harmless amount of energy when a widened beam is used. Furthermore almost the total energy is reflected because the eardrum is covered with aluminium-bronce (the thickness of the laminas is 1µ). Two windows (each approximately 3o ns), whose distance is adjustable, were obtained using an optoelectronic switch (pockels cell). After splitting, the object-beam was directed through the glass-fibers of the 17o°-optic, which was developed especially for ear-examination by WOLF. The laser-beam reflected from the eardrum passes through the lenses of the endoscopic optic and through an additional widening-lens onto the photographic plate. The widened reference-beam impinges on the photoplate at almost the same angle.

The equipment was used in altered form for adjustment. A 5 mW HeNe-laser was used instead of the object-beam (moveable hinged mirror). The photoplate was unclapped and thus the light reflected by the eardrum was directed to a TV-camera. Whereas the endoscopic optic is usually inserted by the examiner, in this case the seated person to be examined must maneuver its head in order to insert the fixed optic as close as possible to the eardrum to obtain the maximal light-intensity. The person can control and correct its movements with help of a TV-monitor. As soon as the adjustment was optimal, the mirror which reflects the HeNe-laser into the object-beam was unhinged and the photoplate was swung into its position. This procedure took only about two seconds. The laser was then activated.

Three persons were tested using the frequencies 15oo cps and 4 kcps. A the lower frequency a similar mode of vibration was observed with a maximum in the posterior portion of the eardrum (stapedial region). There was no registrable movement of the umbo at the given sound pressure level. Similar to the observations on the isolated specimen, a splitting of the fringe pattern (nodular line) could also be detected in vivo at high frequencies (fig. 4).

In our study, it is evident that the modes of vibration obtained with holography on the isolated temporal bone actually correspond to those obtained in vivo. The mode of vibration presented by v. BÉKÉSY can no longer be accepted.

Until now only a small number of the holographs obtained can be used because of the micro-movements. For the clarification of various other questions (i.e. differential-diagnostic value of the procedure or the observation of results of operations such as columella-plastic) a further technical development is necessary.

Fig.2 Investigations on the left human temporal bone post-mortem:

a)
1.2
2.8
4.4
5.9
7.5
8.2
5.1
5.1
3.5
4.4

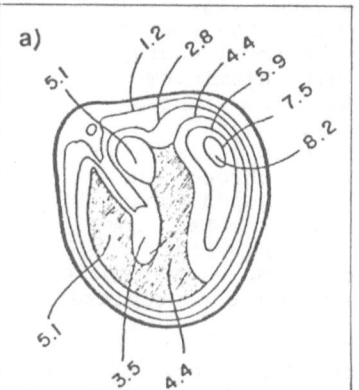

a) TONNDORF and KHANNA, 1972. The figure is slightly turned for better comparison. 525 cps, 121 dB SPL.

b) FRITZE, BURIAN and SCHWOMMA, 1976. An investigation with real-time holography. 2 kcps, 94 dBA.

c) VON BALLY, 1977. Using holography it is not possible to test a range greater than 15 to 20 dB.

~ 1500 nm
b

110dB
105dB
100dB
95dB
C
2,0kHz; 270°/450°

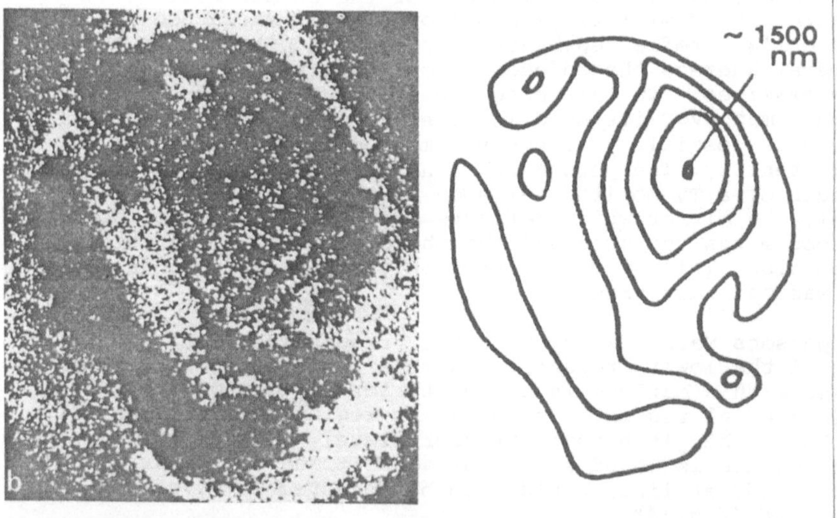

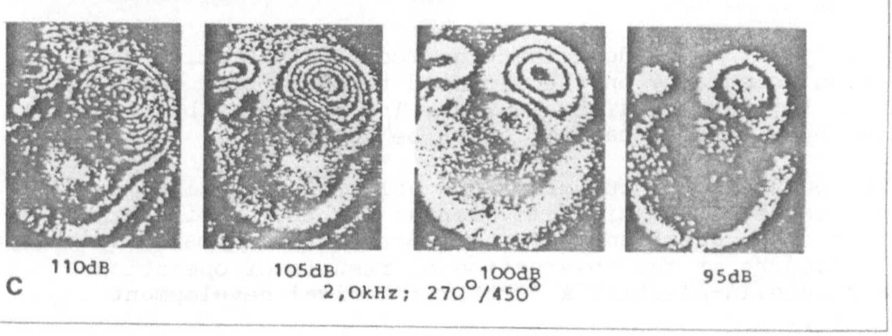

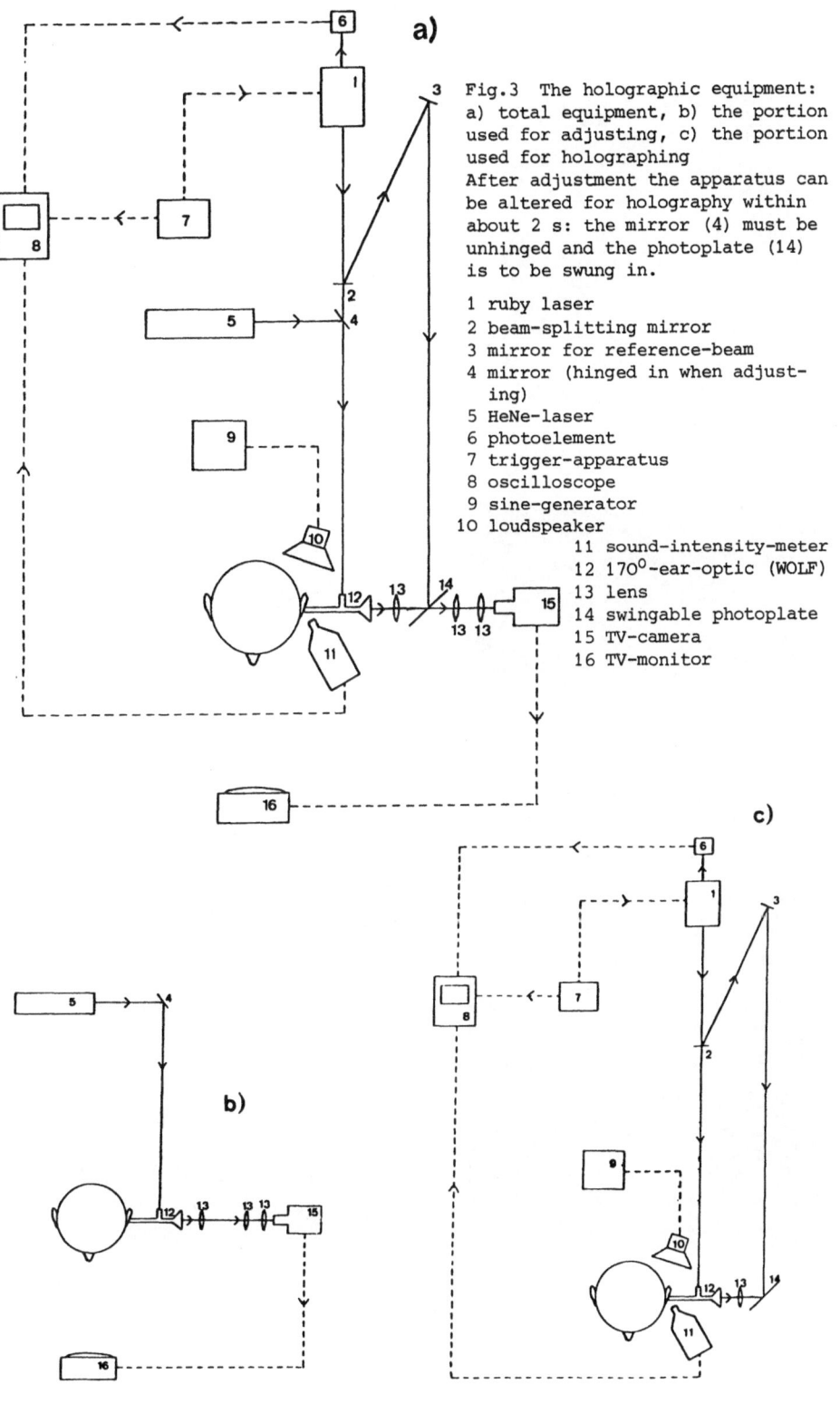

Fig.3 The holographic equipment:
a) total equipment, b) the portion
used for adjusting, c) the portion
used for holographing
After adjustment the apparatus can
be altered for holography within
about 2 s: the mirror (4) must be
unhinged and the photoplate (14)
is to be swung in.

1 ruby laser
2 beam-splitting mirror
3 mirror for reference-beam
4 mirror (hinged in when adjusting)
5 HeNe-laser
6 photoelement
7 trigger-apparatus
8 oscilloscope
9 sine-generator
10 loudspeaker
11 sound-intensity-meter
12 170°-ear-optic (WOLF)
13 lens
14 swingable photoplate
15 TV-camera
16 TV-monitor

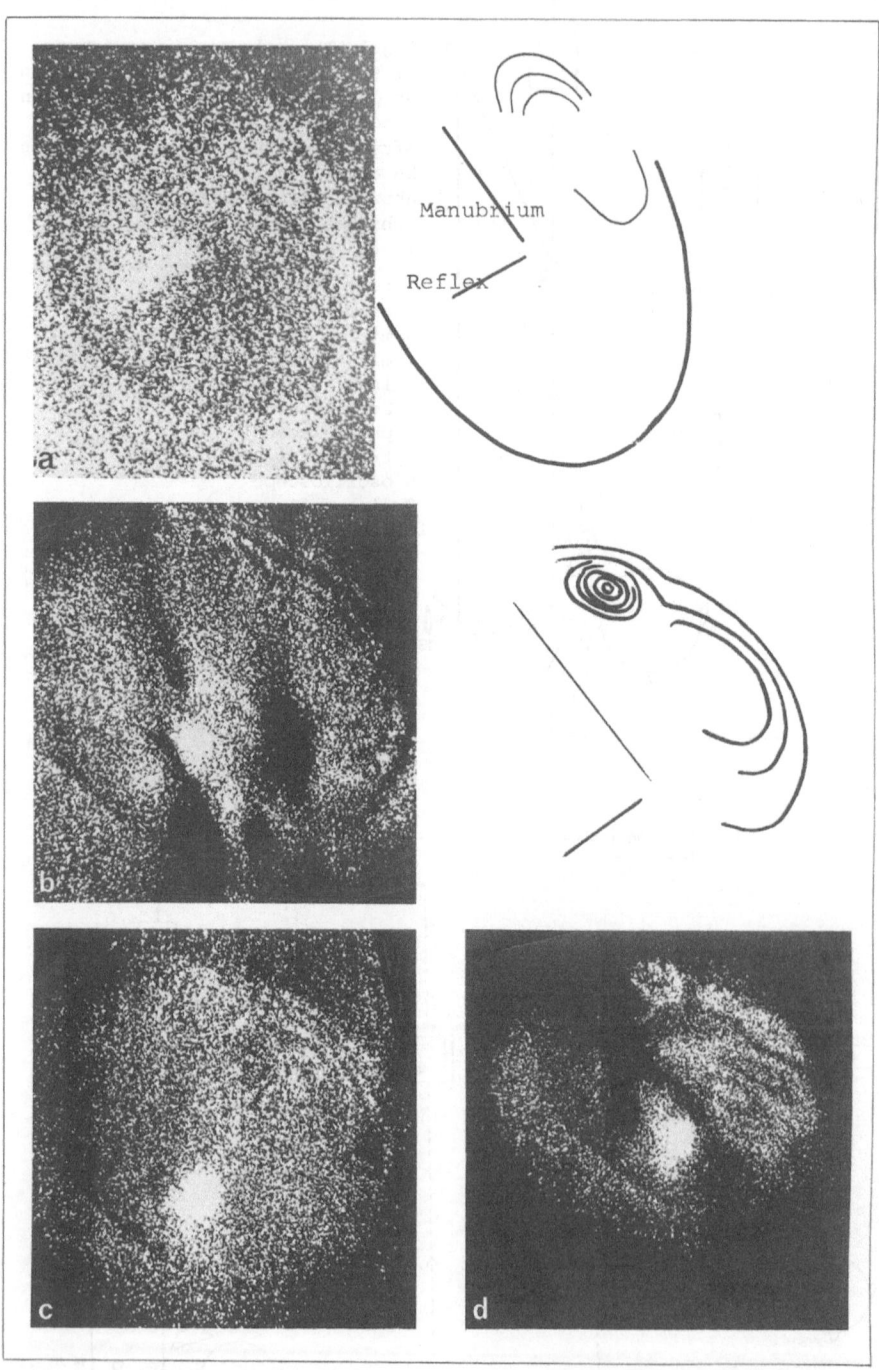

Fig.4 Caption see opposite page

The authors wish to thank the VIENNATONE-Comp. and the WOLF-Comp. for their cooperation.

Literature:

BALLY G.v., 1977
    Holographic Analysis of Tympanic Membrane Vibrations
    in Human Temporal Bone Preparations Using a Double-pulsed
    Ruby Laser System.
    in E. Marom and A.Friesem
    Applications of Holography and Optical Data Processing.
    Pergamon Press
    Oxford and New York

BÉKÉSY G.v., 1941
    Über die Messung der Schwingungsamplitude der Gehör-
    knöchelchen mittels einer kapazitiven Sonde.
    Akustische Zeitschrift 6/1

FRITZE W., K. BURIAN and O. SCHWOMMA, 1976
    Holographische Untersuchung des Trommelfells
    Proceedings of the Symposion 1976
    Special Research Area 88
    Münster

FRITZE W., H. KREITLOW and K. RINGER, 1978
    Holographische Untersuchung der Schwingungsform des
    menschlichen Trommelfells in vivo.
    Arch. Otorhinolaryngol. 221/255

TONNDORF J. and S.M.KHANNA, 1972
    Tvmpanic-Membrane Vibrations in Human Cadaver Ears Studied
    by Time-Averaged Holography.
    J. acoust. Soc. Am. 52/1221

Fig.4 The investigation on living men (left side, $90^0/270^0$)
a) person A.M., 28a; 1.5 kcps, 96 dbA.
   (from FRITZE, KREITLOW and RINGER, 1978)
b) person R.H., 21a; 1.5 kcps, 100 dBA.
c) person G.C., 22a; 1.5 kcps, 100 dBA.
d) person G.C., 22a; 4 kcps, 100 dBA.
   The vibration is splitted just as seen on temporal bones.

Use of ESPI to Measure the Vibration of the Human Eardrum in vivo and Other Biological Movements

O.J. Løkberg, K. Høgmoen[1]
The Norwegian Institute of Technology, Dept. of Physics
N-7034 Trondheim-NTH, Norway and

T. Gundersen
Dept. of Otolaryngology, University of Trondheim
N-7034 Trondheim, Norway

1. Introduction.

We have for some years been working on increasing the application range of electronic speckle pattern interferometry - ESPI - especially for vibration analysis. By introducing a vibrating mirror in the reference path we obtain phasemodulating effects, which have greatly increased the amplitude measuring range from originally below one decade to more than six decades [1-2]. We have also used phasemodulation to separate the amplitude - and phasedistribution across vibrating surfaces [3], - a procedure which is most valuable if compound modes are to be analyzed.

The positive features of ESPI as outlined later make the technique very attractive for studies of biological objects. In close collaboration with Dr. Gundersen we first concentrated on vibration analysis of temporal bone preparations[4-5]. Lately we have extended the technique to measure the vibrations of very unstable objects [6], which has enabled us to measure the vibrations of the human ear-drum in-vivo [7], without resorting to pulsed lasers. Another variation of the ESPI-technique can be used to study more irregular movements of biological objects [6].

2. ESPI Adapted to Living Specimen.

ESPI might be described simply as image holography where the filmregistration has been replaced by the photoelectric action of the TV-camera. The subsequent electronic processing with TV-monitor presentation is equivalent to the holographic reconstruction process. (For a more detailed explanation of the basic principle of ESPI the reader might consult e.g. [8]).

A vibrating object therefore will be shown on the TV-monitor, covered with the same J_0^2-fringes as in holography, if a single exposure is used, while $\cos^2$-fringes will be the result of double-exposures.

Compared to holography, ESPI has a low picture quality due to the limited resolution of the TV-system. This is, however, amply compensated by its advantages:

  real time presentation of interferograms
  short exposure time (normally 40 msec.)
  high repetition rate (25 Hz)

The real-time presentation is very practical in vibration analysis where usually many parameters have to be varied and their effect observed. The combination of short exposure and high repetition rate make it easy to work with relatively unstable objects like ear preparations.

---

[1]Present address: Det Norske Veritas, N-1322 Høvik, Norway

For work on living specimen the stopping action of the ordinary ESPI-system is not sufficient to produce observable fringepatterns of the vibrations. We have therefore shortened the exposure time of each TV-frame by use of a rotating chopper wheel synchronized to the TV-framing. This gives a considerable increase in stability. We might for example hold a vibrating object freely in the hand and still observe its vibration patterns. Provided the exposure is set at an integer number of vibration periods the vibration pattern is given by the usual J_0^2-function. At non-integer values the fringe functions are slightly altered but the phasemodulation techniques described in ref. [1-3] are still effective.

The single (short) exposure technique is best suited to measure vibrations at relatively high frequencies as those excited in the ear by external sound. Most natural biological movements have a slow and irregular nature. If we use a single, short exposure on such a movement, only the parts of the object which moves at low or no velocity will be shown as bright fringes. To study those irregular movements we make use of the fact that each TV-frame in the ESPI-system can be considered to be a properly recorded and reconstructed hologram. Therefore, if we double expose within each TV-frame the resulting ESPI-image will be covered by the usual $\cos^2$-fringe pattern which represents the changes in the object between the exposures.

When we use this technique to study a complex movement, the interference patterns displayed on the TV-monitor change their appearance very rapidly. We therefore record the interesting sequences by use of a videorecorder and replay it afterwards in the singleframe or slow-motion mode whereby the different patterns can be studied.

3. Experiments.

So far we have mainly concentrated on showing that the methods described here can be used on living humans.

The shortened and double exposures have so far been obtained by passing the Ar-laser beam (0.4 W - single mode) through appropriate hole or hole-pairs in a wheel which rotates at the TV-framing rate (25 Hz).

3.1. Vibration Measurements of the Human Eardrum. - Experimental Procedure and Results.

The eardrum was excited by the free sound field from a loudspeaker, while the sound pressure level (SPL) was measured by a microphone close to the ear. The SPL-values cited in this paper must be used with care as they varied quite considerably with the position of the microphone.

The person was placed on a bench next to the ESPI-set up, no strapping or anaesthetizing was used. We used exposures from 2 to 0.5 msec. to reduce the effect of bodily movements upon the vibration fringe patterns.

The most difficult problem was posed by the correct alignment of the illumination and observation directions through the narrow eartube on to the eardrum. We did illuminate and observe the membrane vertically by use of a (variable) beamsplitter as shown on fig. 1. Thus the operator could view the membrane directly in-line with illumination/observation directions. The beams could be rotated and translated as indicated (an additional mirror arrangement is not shown). In addition the subject himself could watch his

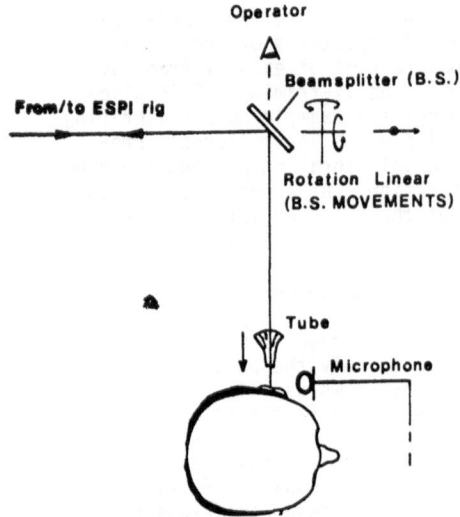

Operator

Beamsplitter (B.S.)

From/to ESPI rig

Rotation Linear
(B.S. MOVEMENTS)

Tube

Microphone

Fig. 1. Illumination/observation of eardrum

eardrum interferograms on a separate TV-monitor and his active cooperation was very advantageous.

The optical properties of the membrane are difficult, as we get large components of internally reflected light in the regions where it is semitransparent. This gave timeaverage fringepatterns of low contrast as we did not want to coat the drum in the preliminary experiments. The contrast problem was solved by use of phasemodulation [1]. The reference wave was modulated with an amplitude that corresponded to a zero point of the J_0^2-function. (The 1.0 and 2.0 zero points correspond to vibration amplitudes of 0.1 μm and 0.22 μm at λ = 0.5145 μm). Then the unwanted scattered light would normally reconstruct with zero intensity. However, the light reflected from the parts of the eardrum that vibrated with the same amplitude and phase as the modulating mirror would reconstruct with maximum intensity representing the shifted zero order fringe. This bright fringe could easily be detected on a dark background as it was moved about on the eardrum by variation of the SPL and/or the modulation. How this works is shown in fig. 2.

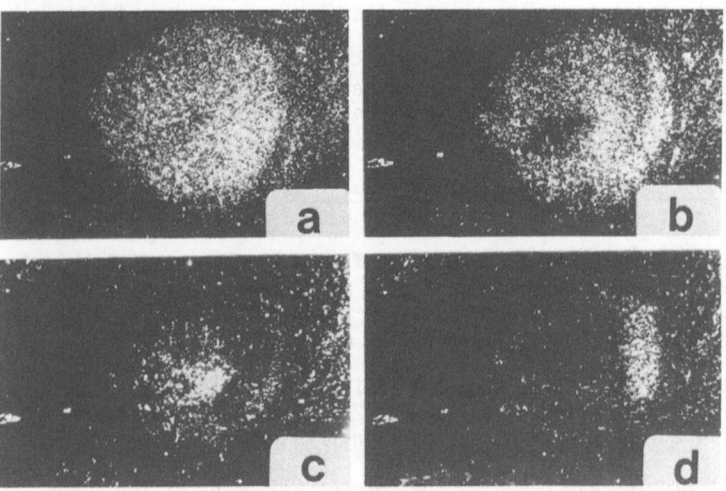

a

b

c

d

Fig. 2a represents the membrane at rest, while fig. 2b-c are recorded at 2400 Hz - 90 dB SPL. Fig. 2b is a time-average interferogram without modulation where it is very hard to make out any positive detection of a vibration pattern. Figs. 2c-d are recorded with phasemodulation at the first zero point (0.1 μm amplitude), but with the modulation phase set at 0° and 160°

which position the zero order fringe at different parts on the membrane. If we modulate the reference mirror at a slightly different frequency [3], the movement is shown very vividly in slow-motion.

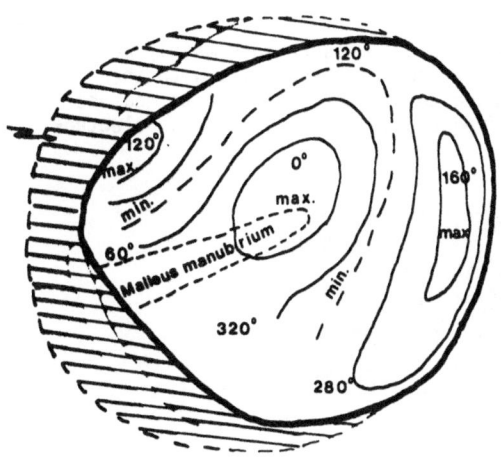

From recordings like those shown in fig. 2, we have sketched in fig. 3, the vibratory behaviour of this particular eardrum at 2400 Hz, as an example (hatched area: not observed parts of the eardrum). The motion is complex with surprisingly great variations in vibration phase. In other drums, however, the movements at the same frequency have been much more uniform with small phase variations.

During the experiments we made extensive use of a videotaperecorder for subsequent analysis of - and measurements on the re- corded interferograms. By means of such videorecordings we could also use the photoelectric method described in [2] to measure

Fig. 3. Eardrum vibration at 2400 Hz.

the amplitude and phase values at low SPL (70-30 dB). In this way vibration amplitudes down to 1 nm could be measured.

By a combination of visual observation and photoelectric measurements we have obtained a kind of frequency response curve of the malleus (hammer) manubrium tip of the left ear of two male subjects shown in fig. 4.

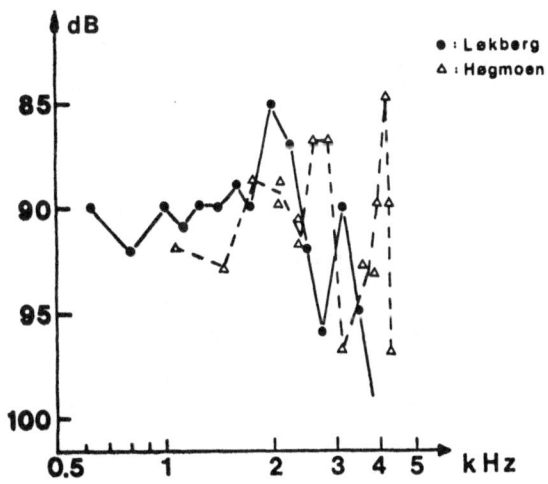

Fig. 4 gives the SPL neces- sary to obtain the first fringe (0.1 μm amplitude) as a function of frequency. The shape of the curve is similar to the curves Gun- dersen [9] observed using an electro-magnetic pick-up.

Both subjects show a point of resonance in the region of 2 kHz. In temporal bone preparations the resonance was between 800 and 1200 Hz. The shifting towards higher frequencies in living man may be due to contractions of the muscles in the mid- dle ear.

Fig. 4. Frequency response of malleus tip

Between 3 and 4.5 kHz there is another peak and the explanation of this may be the increasing of the sound energy in the inner part of the extended meatus (remember; the SPL was measured outside the ear).

3.2. Double-exposure ESPI.

With this technique, we have not made any serious measurements mainly because our rotating chopper is extremely difficult to syncronize with the movement and the TV-framing. It is planned to use the technique to measure the vibrations of the human ear at low frequencies. In that case an electro-optical shutter has to be used.

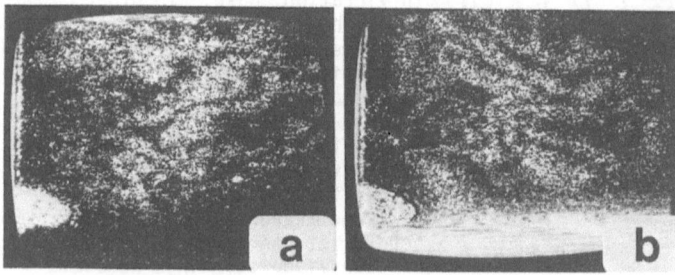

Fig. 5. Fringe pattern of human hand at rest

Fig. 5a-b shows two consecutive TV-frame recordings of the back of a right human hand (the thumb is at the lower left of the pictures). The hand was clamped firmly to a heavy support. The fringes represent the movements of the skin surface in the time interval (3 msec.) between the exposures. When the videorecorder was replayed in the slow-motion mode, the fringepattern seemed to move to and from the fingers (due to bloodpulsation and nervous movements?).

The low quality of the fringepatterns in fig. 5 is mainly due to low light level on the TV-target. In a similar experiment the human heart area [app. 10 x 13 cm^3] was covered with thin, retroreflective tape, whereby more than 20 fringes could be detected across the monitor image.

4. Concluding remarks.

We have shown that ESPI used in conjunction with chopped laserlight can be used to observe and measure induced vibrations and more general movements of living objects. In future work on ear-drum vibrations, the SPL-measurement will be improved, the drum coated to increase the fringe contrast while the malleus head is to be marked for easier identification. Preliminary tests with a new ESPI-technique indicate that the bodily movements can be partly compensated. This will give longer exposures whereby He-Ne lasers can be used as light sources. We are also working on the possibility of measuring the amplitude and phasedistribution from the direct videosignal by means of microcomputer.

The double-exposure technique is really waiting for a repetetively double-pulsed laser which can be synchronized to the TV-framing rate. But while we are waiting, we can chop the light from a cw-laser to observe the movement of small biological object areas. An obvious development is to couple the technique together with fiberoptics to observe the movements of inner organs.

References

1. O.J. Løkberg and K. Høgmoen; J. Phys. E, 9, 847 (1976)

2. K. Høgmoen and O.J. Løkberg, Appl. Opt., 16, 1869 (1977)

3. O.J. Løkberg and K. Høgmoen, Appl. Opt., 15, 2701 (1976)

4. K. Høgmoen and O.J. Løkberg, Proc. Eng. Uses of Coh. Opt., 147, Cambridge Univ. Press (1976)

5. K. Høgmoen and T. Gundersen, Proc. Symp. 76 - Conf. on Electrocochl. and Hologr. in Medicine, Münster(1976)

6. O.J. Løkberg, to be published in Appl. Opt.

7. O.J. Løkberg et.al: Appl. Optics, March (1979)

8. J.N. Butters et al., ch. 6, in Speckle Metrology, ed. R.K. Erf. Academic Press (1978)

9. T. Gundersen, Arch. Otolaryngol, 61, 416 (1972)

Vibration Analysis of the Human Skull and Auditory Ossicles by Holographic Interferometry

Y. Ogura, Y. Masuda, M. Miki, T. Takeda, S. Watanabe, and T. Ogawara
Dept. of Otolaryngology, Okayama University Medical School, Okayama 700, Japan

S. Shibata
Himeji Japan Red Cross Hospital, Section of Otolaryngology, Himeji 670, Japan

T. Uyemura and Y. Yamamoto
Department of Precision Engineering, Faculty of Engineering
University of Tokyo, Tokyo 113, Japan

1. Introduction

Holographic interferometry was applied to medical investigation regarding the vibration analysis of some parts of the human sound conducting system.

Vibration analysis in the fields of medicine and biology has been conducted through optical, electrical and mechanical methods. The optical method excels in allowing an easy observation of a rapid displacement on the vibratory surface without coming in contact with the object. On the other hand, it suffers a drawback in that it is not capable of recording the vibratory amplitude unless the amplitude is visible. In contrast, the electrical condenser method or electro-mechanical pickup method is capable of recording only one point which comes in contact with the element of the device at a time though by far more sensitive compared to the optical method.

Holographic interferometry, above all, may be the most suitable method for vibration analysis in that it features advantages inherent in conventional optical techniques as well as the capability of analyzing vibrations with an accuracy and exactitude on a submicroscopical level as small as the unit of laserlight wavelength.

2. Experiments

Materials used were human skulls and temporal bone specimens containing the auditory ossicles. Test sounds of various frequencies and intensities were applied to these materials for recording vibratory patterns by means of "time-averaged holography" [1]. As light source a helium-neon laser with a wavelength of 6328 A was used.

2.1 The Skull

In regard to an exploration of the mechanism of bone-conduction hearing, a study of skull vibration bears an utmost importance.

Three dry human skulls were used being fixed in the suture lines with adhesive solution, "Allon Alfa". The experiment was made to ascertain the vibratory pattern of the skulls as generated at predetermined resonant frequencies of the skull. As in a clinical bone-conduction audiometry, a sound generator was attached to the temple of the skull specimen unilaterally. The skull was fixed to the base with a magnet stand. A delineation of the skull vibration was made, which showed a pattern symmetrical to the bilateral temporal planes (Figs. 1,2).

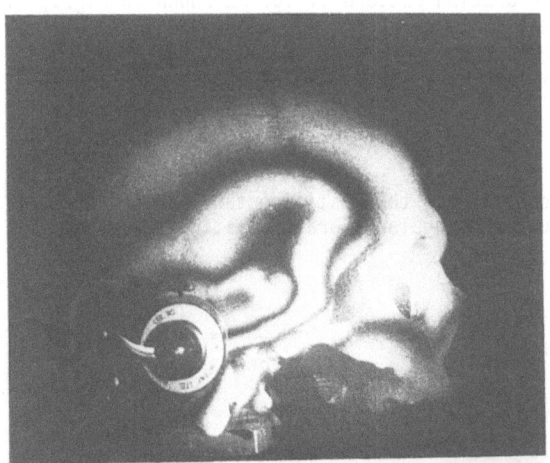

Fig.1 Vibratory
pattern on the right
side of a human skull,
generated at 2,070 Hz -
70 dB

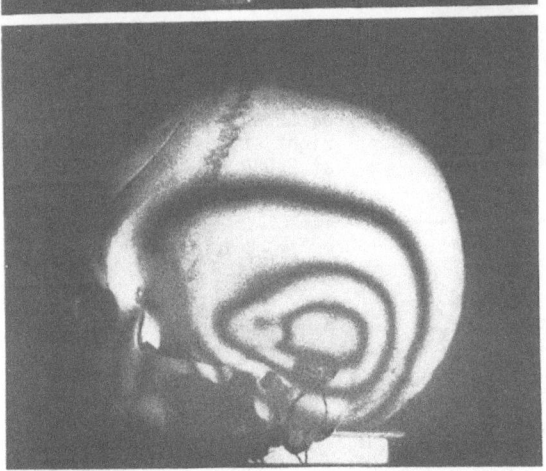

Fig.2 Vibratory
pattern on the left
side of the same human
skull, as in Fig.1, at
2,070 Hz - 70 dB

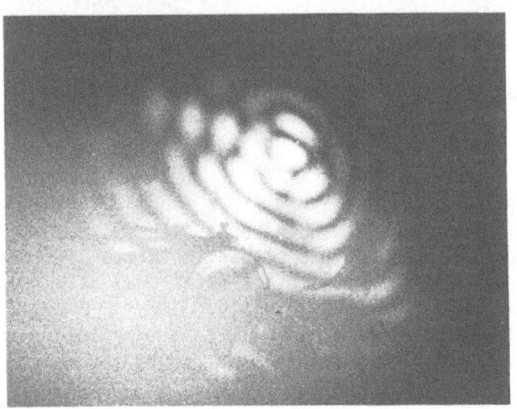

Fig.3 Vibratory
pattern on the right
side of a human skull,
generated at 100 Hz

An interesting finding was the vibration pattern at 100 Hz shown in Fig.3. Those interference fringes mean that a component of the compression bone-conduction might exist even in the lower frequency range as 100 Hz which had been considered to consist of the inertia bone-conduction alone.

2.2 The Auditory Ossicles

The human stapes footplate which constitutes the most important part of the auditory ossicles was studied to obtain singular vibratory patterns.

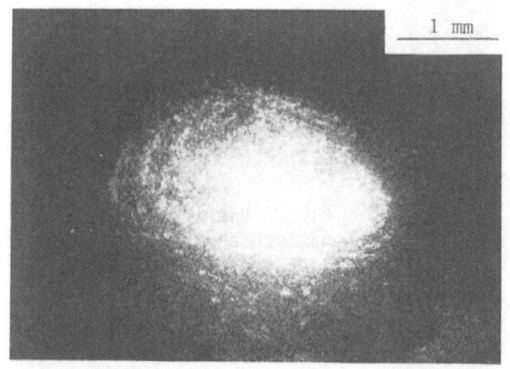

Fig.4 Vibratory pattern of a human right stapes footplate at 2,000 Hz - 90 dB

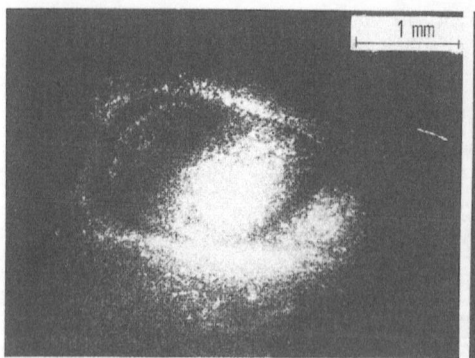
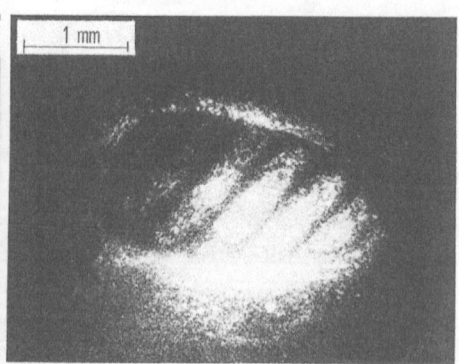

Fig.5 Vibratory pattern of the same human right stapes footplate at 2,000 Hz - 100 dB

Fig.6 Vibratory pattern of the same human right stapes footplate at 2,000 Hz - 110 dB

Figures 4,5 and 6 show holographic reconstructions of a right human stapes footplate as viewed from the internal ear side. As shown here, with increasing intensity of test sounds, the vibratory pattern of the stapes footplate shifted from a piston movement (Fig.4) to a hinge movement with the rotational axis located posterior to the oval window (Figs.5,6). Shown in Fig. 7 is a schematic presentation of stapedial hinge movement rotating around the axis of rotation crossing at 50 degrees with the major axis of the oval window behind the stapes.

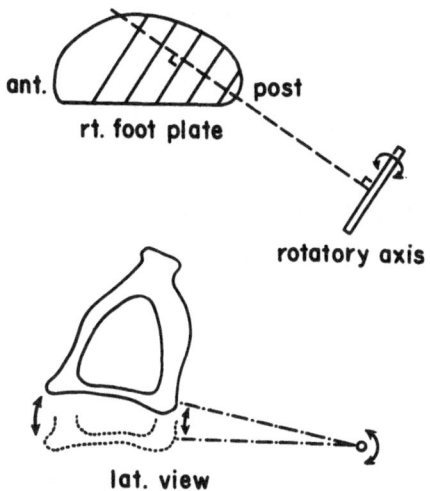

Fig.7 Schema of the vibration mode of the stapes footplate

ant.

rt. foot plate

post

rotatory axis

lat. view

3. Summary and Conclusion

Holographic interferometry has been applied to investigations in the field of medicine and biology[2],[3],[4],[5]. The authors also applied this technique to the study of hearing mechanism and a number of new findings were obtained[6],[7],[8].

As earlier noted, this technique proved to be a unique and rewarding method for vibration analysis capable of performing measurements with an accuracy and exactitude on a submicroscopical level as small as the wavelength of the laserlight.

Nevertheless, holographic interferometry remains a technique which calls for due consideration when it is used in the field of medicine. To give an example, using this technique, the specimens must be prepared in such a way that the parts under consideration may readily be observable and the physical conditions must be kept constant during the whole course of the experiment since temperature, dryness, etc. rapidly change the physical properties of the materials. Countermeasures to cope with these problems thus far developed include thin coating of bronze powder on the surface of the vibrating body such as the tympanic membrane for improved reflection of laserlight and the use of a fresh specimen from an experimental animal or a frozen specimen of the temporal bone taken from a cadaver as fresh as possible.

Hopefully, holographic interferometry will serve itself as a more valuable tool for vibratory analysis in the field of medicine as well as engineering through further improvement and modifications.

4. References

1 Powell, R.L. and Stetson, K.A.: Interferometric vibration analysis by Wave front reconstruction. J Opt Soc Amer 55:1593-1598, 1965.
2 Tonndorf, J. and Khanna, S.M.: The tympanic membrane as a part of the middle ear transformer. Acta Otolaryngol 71:177-180, 1971.
3 Tonndorf, J. and Khanna, S.M.: Tympanic-membrane vibrations in human cadaver ears studied by time-averaged holography. J Acoust Soc Amer 52:1221-1233, 1972.

4 Gundersen, T. and Høgmoen, K.: Holographic vibration analysis of the ossicular chain. Acta Otolaryngol 82:16-25, 1976.
5 Bally, von G.: Holographic analysis of tympanic membrane vibrations in human temporal bone preparations using a double pulsed ruby laser system. Marom, E. and Friesem, A.A. (Ed.): Applications of holography and optical data processing, pp. 593-602, Pergamon Press, Oxford and New York, 1977.
6 Ogura, Y. et al.: A holographic study of the eardrum vibration. Jibi-inkoka(Otologia Tokyo) 46:83-88, 1974.
7 Ogura, Y. et al.: A holographic study of the human skull vibration. Audiol Jap 19:163-167, 1976.
8 Masuda, Y. et al.: A holographic study of the human stapes footplate vibration. Audiol Jap 20:709-712, 1977.

X. Acoustical Holography

State of the Art and Future of Acoustical Holography in Medicine and Biology

P. Greguss

Applied Biophysics Laboratory, Technical University Budapest
Budapest, Hungary

The first acoustic holograms were recorded just 15 years ago, and a few months later even ultrasonic holograms from a living human eye having a tumor were reconstructed [1,2]. In the past one and a half decade several attempts have been made all around the world to use ultrasonic holographic imaging in medical diagnosis, nevertheless, existing ultrasonic imaging methods yield under clinical conditions a far better image quality at less cost. The scope of this review is to investigate whether ultrasonic holography can ever produce anything approaching a realistic 3-D reconstruction of insonified objects in the frequency range used in medical diagnostics, and how acoustical holography can help differential diagnostics when not the knowledge about shape but the knowledge about structure is important.

The Feasibility of Acoustic Holography

A hologram per definition is no more than an interference pattern originating, for instance, from the interaction of the wavefront scattered from the object, $F(x,y)$, and a coherent reference background $\exp(-kr)$, and it is obvious that such a pattern can be produced during insonification. If the optical replica of this acoustic interference pattern, i.e., the transparency of it, has an amplitude transmission proportional to the intensity of the two interfering acoustic waves, then, when illuminated with coherent light, one obtaines in the plane behind it an optical amplitude and phase distribution $H(x,y)$ which contains beside some other additional terms a term equal to the phase and amplitude distribution of the object wave $F(x,y)$:

$$H(x,y) = (1 + |F|^2)\exp[i(k_1 x + k_2 y)] + F(x,y) +$$
$$+ F^*(x,y)\exp[i2(k_1 x + k_2 y)] \qquad (1)$$

In the sense of Huygens' principle an optical wave develops behind the transparency and, due to the second term on the right hand side of (1), a sound image could be visualized in space.

Two factors which significantly affect the quality of the re-constructed image are the size and spatial frequency response of the recording medium. The size of the hologram recording surface determines the image resolution and depth of field, whereas its spatial frequency response determines the angular field of view and the image intensity response. This, however, means that the method and technique by which the acoustic holo-gram pattern is converted into the demagnified optical counter-part can be of fundamental importance.

The reason for introducing a size reduction factor is that due to the difference in wavelength between the insonifying ob-ject wave λ_S and the reconstructing light λ_L the reconstructed wavefront appears to come from an object field with unchanged lateral (x,y) dimensions, but with depth distortion proportion-al to λ_S/λ_L. A completely realistic 3-D reconstruction can, however, be obtained only if the lateral and longitudinal magni-fications are equal. The lateral magnification can be expressed as

$$M_{lat} = (1/m) \; (\lambda_L/\lambda_S) \; (D_2/D_1) \tag{2}$$

and the longitudinal magnification as

$$M_{long} = (\lambda_S/\lambda_L) M_{lat}^{\;2} \tag{3}$$

A three-dimensional sound image can, therefore, be visualized without distortion only if the lateral magnification equals λ_L/λ_S. In the most common domain of sound imaging, the ratio λ_L/λ_S has a magnitude of about 10^{-3} thus the undistorted image will be very small and so one has to use optical magnification to obtain useful images thereby regenerating the depth distor-tion.

Since recorded acoustic holograms have to be scaled down by the ratio of the wavelengths, the visualized sound image does not look three-dimensional because demagnified acoustic holo-grams in the majority of cases do not present the viewer a large enough aperture (window) to allow useful parallax. For real three-dimensional impression, at least a "window" of 10 cm is needed, which means that if the acoustic hologram could be taken at a wavelength of $\lambda_S = 10^{-1}$cm, and green light of wave-length $\lambda_L = 0,5.10^{-4}$ cm is used for reconstruction, we have to start with an acoustic hologram recording of 200 m (!) which is naturally unrealistic.

The position of the reconstructed and visualized sound image is determined by (4):

$$\frac{1}{\lambda_L D_2} \mp \frac{1}{\lambda_S m^2 D_1} + \frac{1}{\lambda_S m^2 R_1} + \frac{1}{\lambda_L R_2} = 0 \tag{4}$$

where λ_S and λ_L are acoustic and laser beam wavelengths, D_1 and D_2 are object and image distances, R_1 and R_2 are the radii of curvature of the reference and reconstruction beams, and m is the size reduction factor introduced in the recording of the hologram.

As seen from this brief review on the principles of holography, it is evident that the feasibility of acoustical holography exists, i.e., 3-D information is there, but it is available only one plane at a time, i.e., one depth plane after another can be brought into focus on a screen or vidicon surface. This type of imaging called "tomographic imaging" is, however, not related to conventional X-ray imaging, where the information of several planes is compressed in a single plane, but it is equivalent to ultrasonic C-mode imaging obtained by time gating, and it can also be regarded as a relative to ultrasonic B-mode images.

Problems of Recording

The quality of the visually reconstructed acoustic image is strongly influenced by the number of fringes in the recorded Fresnel zone pattern which, however, is a function of the geometric size of the recording medium. In this aspect the sonosensitized plates - on which the first ultrasonic holograms were recorded - seem to be ideal, since areas of several hundreds of cm^2 can easily be achieved. However, a serious drawback of sonosensitized plates is that besides having a low "speed" (i.e., their response to low ultrasonic intensities is poor) their threshold contrast hardly meets the values needed for an adequate reconstruction.

From the point of view of the size of the recording medium acoustical-to-optical area detectors based on nematic liquid crystals would also be adequate, but we are here faced with the same problems as with sonosensitized plates: low sensitivity, low threshold contrast.

Since to record a hologram which allows a rather good reconstruction, the diameter of the recording area has to be in the order of 100 wavelengths, and the face plate of Sokolov-type ultrasonic vidicon has this dimension, it seems to be a good candidate for medical ultrasonic holographic imaging: it works in real time and it has a very high "speed". However, the area of the piezoelectric faceplate is to be regarded as a closely packed array of receivers with apertures equal to approximately 1,5 times the plate thickness, and since the wavelength in the plate is four times larger than in the surrounding water, the angular field of view is restricted to less than 18°. If, however, the piezoelectric plate is replaced by an electret array, it appears that angular apertures up to 60° are feasible without appreciable loss in sensitivity. Since in contrast to Sokolov type ultrasonic vidicons with piezoelectric face plate, the thickness of the back of the electret face plate is essentially separated from the parameters that determine the sensitivity of the electret elements, physical apertures larger than 100λ can be designed without influencing the uniformity in sensitivity across the face plate.

At present the most frequently used and advertised techniques for recording ultrasonic holograms are the so-called liquid surface deformation methods. Although the area of the recording surface is significantly smaller than in the previous ones, the threshold contrast between pixels on the recording surface is far better. The liquid surface levitation methods can be classified in three groups:

a) the deformation occurring at frequency of the acoustic wave,

b) the deformation is due to radiation pressure against gravity, which produces a vertical displacement of the total ripple pattern having a natural frequency of about 25 Hz,

c) the deformation which produces the hologram is also due to the radiation pressure, but acts against the surface tension. It has a natural frequency which is proportional to the 3-halves power of the spatial frequency of the ripple, which is generally in the order of 2500 Hz.

The best hologram recording result is achieved when insonification with about 10^{-3} - 3.10^{-4} sec ratio is used. Therefore, to visualize the reconstructed acoustic hologram light flashes in the order of 5.10^{-6} seconds are needed.

The ultrasonic holographic recording methods discussed so far use area detectors, however, an ultrasonic hologram pattern can be recorded also by one of the methods by which sampled sound images are formed. One of the simplest ways is when a point like receiver mechanically scans over the region of intersection of the object and reference beams, the signal of which modulates a point source in synchronism with the scanning motion of the aperture and the position - intensity history of the lamp which in fact is the visualized acoustic hologram is then recorded by a camera. An advantage of this method is that no effective ultrasound reference beam is needed to record the holograms, it is enough to record the scattered ultrasonic field of the insonified object and to simulate the reference wavefront by adding or multiplying the received electric signal with a reference electric signal. Multiplication is preferred since it produces the desired signal free of the extra unwanted components that result from the additional process.

Scanned holographic systems are normally operated in reflection mode and use range gating to eliminate the effect of unwanted reflections. A resolution improvement by a factor of 2 can be achieved for a given size scan aperture by incorporating the insonifying transducer and the receiving aperture on a time share basis in a single unit.

The sampling process always introduces some unavoidable data degradation. This could be tolerated, however, the recording time is so long that it prohibits recording of insonified objects in motion, thus preventing most types of medical applications.

A scanning speed that allows real-time operation can be achieved, according to a suggestion of Berbekar [3], by sweeping the frequency of the insonifying sound beam. This concept is based on the premises that a hologram or the Fourier spectrum of an object contains information on unimportant details and so in most cases of acoustic holography it is enough to pick up only predetermined parts of the hologram, and so the number of receivers or processing time can be decreased.

A detector moving along the diffraction pattern with a constant velocity v will experience an intensity variation of

$$I(t) = \frac{k}{\lambda} \left[\frac{\sin(\frac{2\pi.d}{2\lambda_o} v.t)}{\frac{2\pi.d}{Z\lambda_o} v.t} \right]^2 \tag{5}$$

where Z is the distance from the slit having a width d, and k is a constant depending on the amplitude of the insonification and the geometry of the arrangement.

However, if the detector is fixed and the frequency is varied with the sweeping rate

$$m = \Delta N/\Delta T \tag{6}$$

where N is the frequency of the wave, the detected intensity as a function of time will be

$$I = \frac{k}{c} m.t \left[\frac{\sin(\frac{2\pi d}{cZ} x_o m.t)}{\frac{2\pi d}{cZ} x_o m.t} \right]^2 \tag{7}$$

In both cases, the sweep of the detected signals will be similar if there is a correlation between the moving velocity of the detector with constant frequency and the speed of frequency sweeps according to

$$v = \frac{m \, x_o}{N_o} \tag{8}$$

The only difference is that the height of the maximum increases with frequency, due to the m.t factor before the brackets in (7).

Although by increasing frequency the distance between the minima in the diffraction pattern decreases, the intensity of the fringes is constant. Therefore, it should be compensated for by decreasing the intensity of the insonifying beam in a suitable way. Fortunately, however, in medical ultrasonics there is a "natural" compensation for it, because for most tissues sound wave attenuation increases approximately in the same way as the distance between the minima decreases.

A great advantage of this scanning technique is not only that no moving parts are needed, but also that rather simple computer can be used for image reconstruction. Since ultrasonic sweep generators and wide band transducers are commercially available, one really wonders why this technique does not find medical applications.

A holographic recording method, which has a great resemblance to the B-mode technique used in ultrasonic diagnostics, explores the fact that a signal issuing from a point reflector located on the y axis at $y = R_o$ has the form at x

$$A_r = kA_o\cos\omega(t + \frac{2r}{c}) \tag{9}$$

if the from x transmitted wave has had the form of

$$A_T = A_o\cos\omega t \tag{10}$$

where ω is the angular frequency, c the sound velocity in the propagation medium, and k a constant.

If a coherent reference signal is added, then, after filtering, the form of the detected signal will be

$$A_D = k'A_o\cos\left[\frac{4\pi}{\lambda}(x^2 + R_o^2)^{1/2}\right] \tag{11}$$

This is, however, the equation of a one-dimensional Fresnel zone plate pattern which over a fairly large range of x is a sinusoid whose frequency is a linear function of x for any particular value of R_o. At any value of R_o the synthetic aperture signal for a real target will consist of an array of overlapping Fresnel zone functions. To reconstruct the image of the target at range R_o, the cross correlation integral has to be applied, using the function for a single point target at that R_o for one of the functions in the integral and the synthetic aperture signal for the other. This procedure can be performed either by a digital computer or by an optical correlator. In the latter case holographic film is used to record the one-dimensional Fresnel zone pattern.

Since this recording method can be regarded as the operation of a physical one-dimensional linear array with a large effective transducer aperture, it is also called "synthetic aperture" technique. In order to use it for ultrasonic medical purposes the problems of conflicting requirements concerning the transducers for B-mode and holographic imaging have to be solved. Good hologram recording namely requires strongly focused, short focal length transducers with long pulses, while B scans need weakly focused transducers with long focal length and short pulses. As shown by several investigators [4], these requirements can more or less be met in ultrasonic diagnostics, and since only 23 seconds are needed to scan, for instance, a hologram aperture of the eye, ophthalmology may become one of the rare medical fields where acoustical holographic imaging could have a future.

Problems of reconstruction

Already from this brief and far not complete review of ultrasonic holographic recording techniques it comes clearly to sight that the crucial problems of ultrasonic holography are in realty not associated with recording, but most probably in achieving an undistorted visible reconstruction of the insonified scene which then can be evaluated meaningfully. So, e.g., the left side image of Fig. 1 shows a tumor implanted under the skin of a rat as photographed from the reconstruction of an ultrasonic hologram recorded on a sonosensitized plate in 1965 [5] , while at the right side another tumor in a skin flap can be seen as photographed from the reconstruction of an acoustic hologram recorded by surface levitation method in 1966 [6]. These two images are reconstructions from insonified soft tissues, but the overall quality of the images of mixed structures is also about the same, as shown in Fig.2, which is an acoustic holographic image of a

 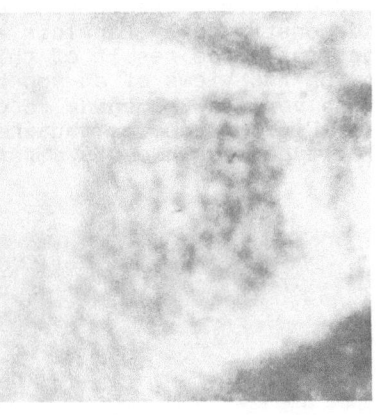

Fig.1 Ultrasonic hologram reconstructions of implanted tumors recorded on sonosensitive plate (left) and by surface levitation method (right)

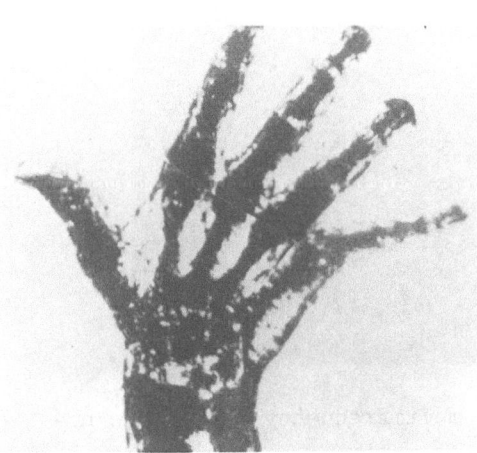

Fig.2 Holographic image of a hand recorded by the acoustical holographic system of Holosonics, Inc.

hand recorded by the acoustical holographic system of Holosonics, Inc. in 1972. The rather poor quality of these images as compared to those obtainable by B scans is, however, essentially not the result of the holographic method itself but rather the difficult-to-avoid consequence of the fact that insonification with ultrasound in general means a highly coherent "illumination" of the target. With other words, the amplitude of the scattered coherent wavelets sent into the picture from the volume outside the focus region are summed vectorially and then squared, i.e., it is not the intensities that are summed as if incoherent insonification would be used, resulting in a uniform "gray" background. This phenomenon called speckle noise results in an intensity fluctuation which can be as high as 1:10 000, and is responsible for the reduction in resolving power by at least one or two orders of magnitude.

Several suggestions have been made in order to get rid of these disturbing effects. One way is to use a digital storage device, a PEP Scan Converter, and then store and integrate 50 images reconstructed from a sequence of acoustic holograms recorded at 50 different frequencies evenly spaced between 2 and 3 MHz, as demonstrated by Langlois at Holosonics, Inc. [7]. Fig.3 demonstrates the result of this rather complicated technique, showing two views of an upper arm with vascular structure. I think it is very enlightening to compare these sound images with that of Fig.4 which was obtained by photographing the screen of a Sokolov typetube which is a far simpler US imaging device.

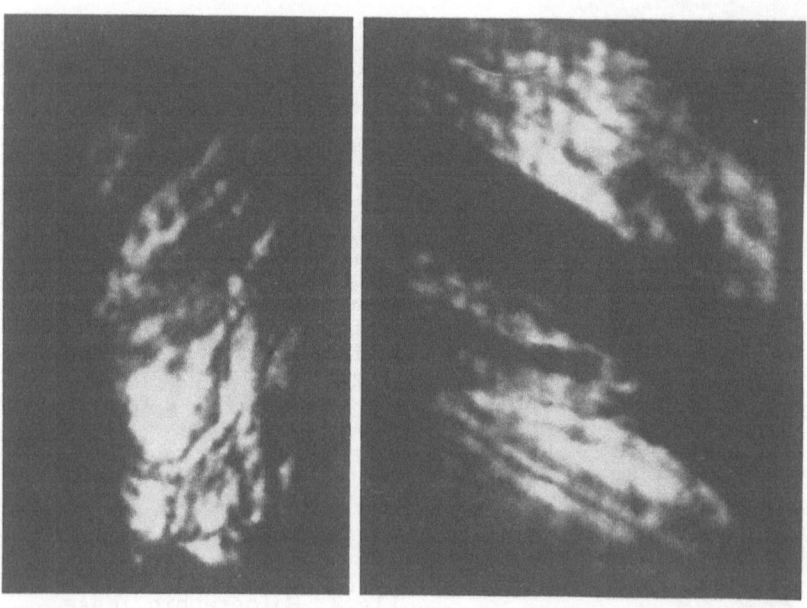

<u>Fig.3</u> Reconstruction of composed multifrequency ultrasonic holograms

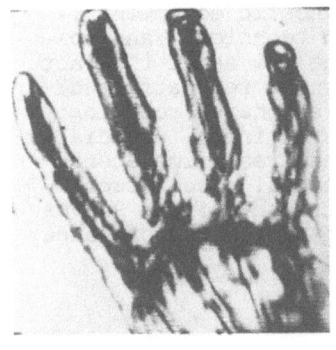

Fig.4 Sonopicture of a hand photographed from the screen of a Sokolov type vidicon tube

Fig.5 Sound images reconstructed from ultrasonic holograms recorded by pulsed sound and pulsed laser illumination

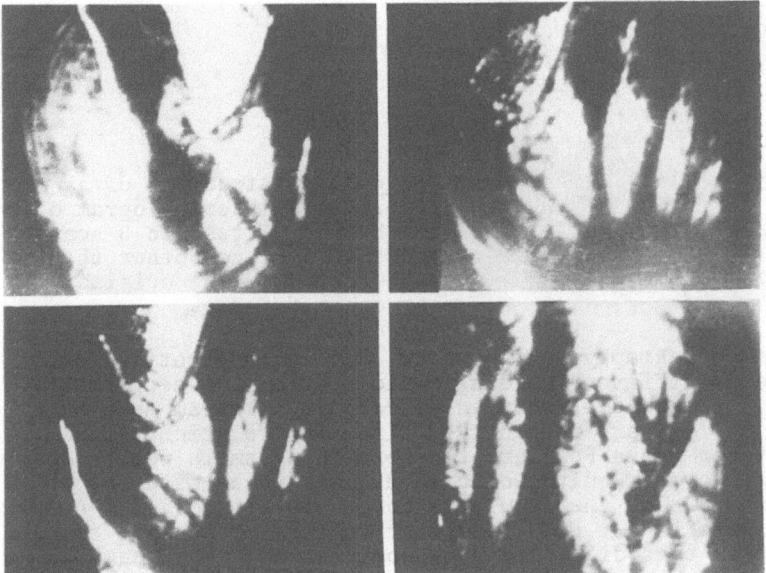

Another way to reduce the image degradating effects is to use pulsed sound and pulsed laser illumination. The sound images of Fig.5 are reconstructed from this type of holograms recorded by Shae at the Shanghai Institute of Cell Biology of Academia Sinica. The right hand pictures are different views from a palm, the left hand ones from an upper arm.

Has Acoustic Holography a Future?

When comparing these medical ultrasonic images with up-to-date B-mode images, we have to draw the conclusion that acoustic holographic imaging - at least in the frequency range up to about 20 MHz - yields image qualities inferior to those already ob-

tainable with less expensive ultrasonic diagnostic equipment .
The reason for this is more <u>fundamental</u> than technical, and so
too much improvement cannot be expected. But how about the fact
that from a single hologram several views can be reconstructed?
With the appearance of real-time ultrasonic scanners (rotating
or array), however, these virtual advantages diminish, especially
if laser holographic multiplexing of different real-time scenes
is considered [8]. (Fig. 6). Further, in several cases, recording a laser hologram from the visualized ultrasonic field solves
this problem, too. The ultrasonic scattering properties of bones
have been studied using this technique [9] .

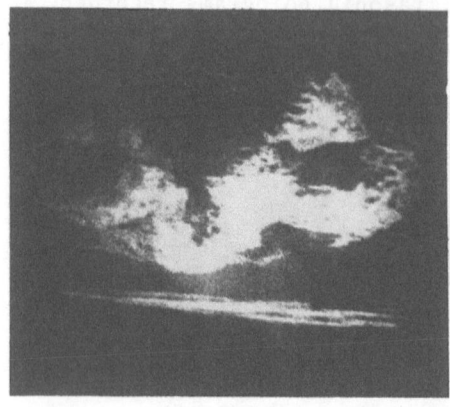

<u>Fig.6</u> Reconstruction of a
multiplexed laser hologram of
ophthalmic ultrasonic B scans
shows also planes other than
represented by the original
sequence of images

Ultrasonic holography may, however, become a versatile tool
in solving some differential diagnostic problems. One possibility is, e.g., to use the recorded ultrasonic hologram as a complex filter for pattern recognition since the relation between
a complex filter and a Fourier transform hologram is well known.
In this case, however, ultrasonic waves have to be used for reconstruction and then the result is to be visualized. The problem in this case is not the visualization itself, but how to record a hologram that diffracts ultrasonic waves. The feasibility of this idea has already been proved for nondestructive
testing, and why not to check it for medical diagnosis?

The fact that a Fourier hologram of an extended source of
arbitrary spectral intensity distribution can be recorded with
the aid of a two-beam interferometer with tilted mirrors strongly suggests that ultrasonic holographic Fourier spectroscopy
could be developed to solve some of the soft tissue differentiation problems. Since here the visualization of the insonified
space is not required, the wavelength discrepancy creates only
problems similar to those in IR spectroscopy, and they can also
be solved in a similar way.

As for conclusion: ultrasonic holography helped us in the
past to get a better understanding of acoustic imaging [10] but
it has a restricted future in clinical ultrasonic diagnostics.
However, as a research tool, especially for biologists, it will
serve for quite a while.

References

1. Greguss, P., Research Film 5, 330 (1965)
2. Bertenyi, A., Greguss, P., Proc. SIDUO II, 133 (1967)
3. Berbekar, G., Tökes, S., Ultrasonics 16, 251 (1978)
4. Chivers, R.C., Ultrasonics 12, 209 (1977)
5. Greguss, P., XII. Seminarium z Akustiki, Warszawa 1965)
6. Weiss, L., Holyoke, E.D., Surg.Gynecol.Obst. 128, 953 (1969)
7. Langlois, G., Priv. Comm. 28.9.1977
8. Greguss, P., Caulfield, J., Science 177, 422 (1972)
9. Greguss, P., Ann. N.Y. Acad. Sci. 267, 312 (1976)
10. Greguss, P., Seen by Sound - Acoustic Imaging. Focal Press London (In Press)

XI.   Special Holographic Techniques

Sandwich Holography and Its Applicability to Biomedical Investigations

N. Abramson

Div. Production Engineering, Royal Institute of Technology
S-100 44 Stockholm 70, Sweden

1.   Introduction

Holographic interferometry is becoming more and more used for mechanical testing. When the conventional double exposure method is used the following practical evaluation problems have to be solved:

The sign of the displacement cannot be found because the hologram contains no information on which exposure was made first.

If many different situations are to be compared, e.g. during a step-by-step loading of the object, a new hologram has to be exposed before and after each load change that is to be studied. New combinations cannot be made during reconstruction.

When the load is applied, unwanted motion of the whole object and its fixture very often complicate the fringe pattern in such a way that the wanted information is hidden.

If the deformation of only one part of the object is to be studied the evaluation is complicated by fringes caused by the total object deformation.

The stress and strain caused by bending is represented by the change of fringe separation over the total object surface, thus the derivative of the fringe separation has to be calculated, which might cause difficulties.

The ordinary fringes represent digital information, fractions of a fringe are very difficult to measure which results in low accuracy especially if only a few fringes are seen.

If object shape is to be measured by holographic contouring the direction of the intersecting surfaces becomes fixed during exposure and cannot be changed during reconstruction. A rotation of the intersecting planes during reconstruction would be useful, e.g. for checking the parallelism of different object surfaces.

It appears as if the sandwich hologram [1] represents the solution to all these different problems.

2. Making the Sandwich Hologram

One hologram plate was placed in a special holder equipped with three contact points that locate the plate surface while three pins locate two sides of the plate. A second hologram plate was then placed in contact with the first, covering its surface. Both plates had their emulsions towards the object (Fig.1). The emulsions were therefore separated by the glass base of the front plate. When the plates were exposed simultaneously both the object beam and the reference beam reaching the back plate passed through the front plate which was not covered by any anti-halo layer. The two exposed plates were taken away and two new plates were placed in the plate holder.

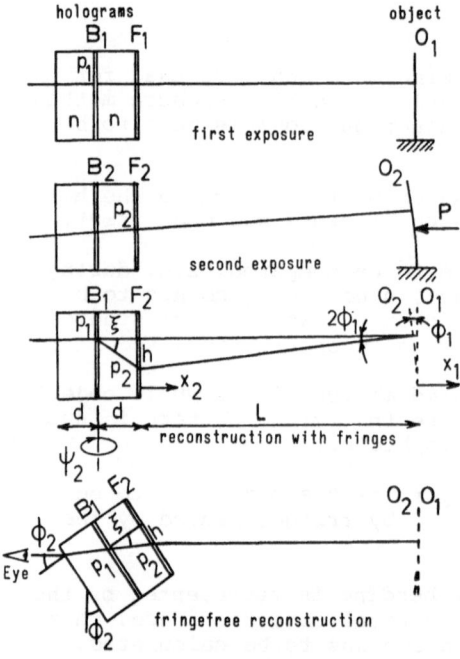

Fig.1. The top of the object O is tilted an angle ϕ_1 by the force P. Hence a speckle ray from one object point is moved the vertical distance h from p_1 to p_2. B and F are the emulsions of back and front hologram plates respectively. Glass plate thickness is d and refractive index is n. ϕ_2 and ψ_2 represent sandwich rotation around a horizontal and a vertical axis respectively. The identical reference and reconstruction beams are excluded from the figure

The object consisted of three vertical steel bars (Fig.2) that were fixed by screws at their lower ends to a rigid frame which surrounded the bars and was supposed to function as an undeformed

236

Fig.2. Rigid motion of the total object has hidden the information concerning deformations. If this hologram had been an ordinary double exposure it would have been judged a failure

fixed reference surface. The middle bar was also supported at the top end. A force was applied to the middle of the bars. After the deformation of the three objects the second pair of plates was exposed. Thereafter all four plates were processed and the back plate of the first exposure (B_1 of Fig.1) was repositioned in the plate holder behind the front plate of the second exposure (F_2). The two plates were bonded together to form a sandwich hologram and reconstructed by a laser beam the direction and divergence of which were identical to those of the reference beam during exposure.

3. Elimination of Spurious Fringes

Strange looking fringes were seen on all three bars (Fig.2). It was very difficult to find any correlation between this fringe pattern and the expected deformations. Straight, inclined fringes that covered the supposedly fixed frame disclosed the situation. The total object, including the frame, had made unwanted rigid motion. Even after this fact had been found it was difficult to evaluate the fringes on the steel bars. If the hologram had been an ordinary double exposure it would simply have been considered a failure.

The sandwich hologram, however, has the property that its fringes can be manipulated by tilting it in relation to the reconstruction beam. By holding the sandwich by hand and tilting it in different directions it was easy to find the reconstruction angle that made the rigid frame appear fringe free (Fig.3). The fringe pattern on the three bars now represented the deformation of the bars in relation to the frame and correlated well to what could be expected from their preknown load situation. Thus we

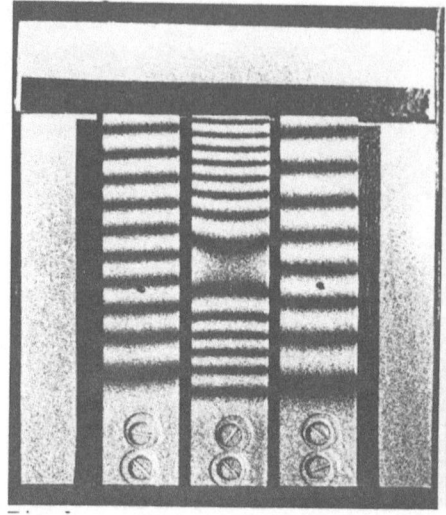 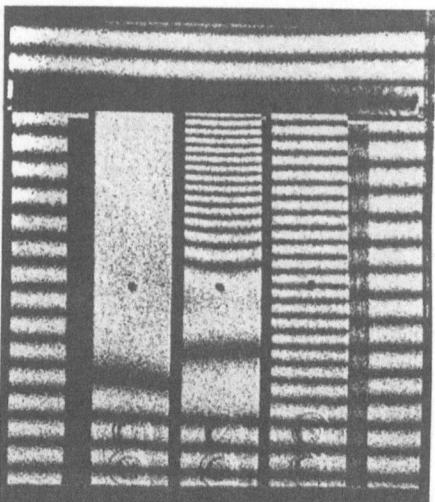

Fig.3.   Fig.4.

Fig.3. The same reconstruction as that of Fig.2 but a small tilt
of the sandwich hologram has eliminated the fringes on the rigid
frame. The fringe pattern now correlates well with the expected
deformations. The object consists of three bars, the lower ends
of which are screwed to a rigid frame. The middle bar is also
supported at the top end. Between the two exposures forces were
applied at the middle of the bars. The right bar is deflected
away from the observer. The other two bars are deflected towards
the observer

Fig.4. The same reconstruction as that of Fig.3 but the sandwich hologram has been tilted so that its top moved towards the
observer. At a certain tilt angle (ϕ_2) the top of the left bar
(which was deflected with an angle ϕ_1 towards the hologram) is
fringe free

had now changed our failure into something that would usually
be considered a successful hologram which any holographer would
be content with. A practical application of this technique is
shown in Figs.10 and 11 where unwanted body motion is virtually
eliminated by tilting the sandwich.

4.   Sign and Magnitude of Object Tilt

But a lot of information is still missing. From Fig.3 we cannot
find out which bar was bent forwards or backwards. What has
happened to the bar in the middle? Should we add all fringes from
the top to the bottom to find the deflection? Or should we let
the fringe number change sign halfway? By tilting the sandwich
hologram so that its top moved towards the observer the number
of fringes on the left bar decreased while the number on the right

bar increased (Fig.4). At a certain sandwich tilt angle the top
of the left bar became totally fringe free. The rigid frame now
was covered by fringes having the same spacing and direction as
those that had been eliminated from the left bar. From this
rather large sandwich tilt the much smaller tilt angle of the
bar could be calculated.

As both magnitude and sign of sandwich tilt are analogous to
object tilt we could conclude that the left bar had tilted towards the hologram while the right bar had tilted away from it.
The number of fringes on the lower part of the middle bar decreased while those on the top increased thus proving that its
lower part had tilted towards the hologram while its top had
tilted in the opposite direction.

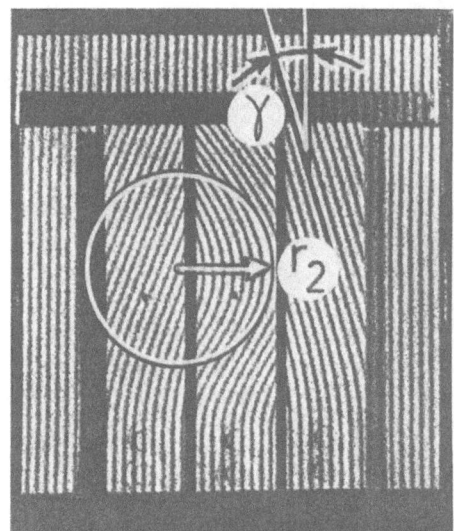

Fig.5. The same reconstruction as that of Fig.3 but
the sandwich has been rotated through the angle ψ_2
around a vertical axis. The
angle (γ) between the fringes and a vertical line is
a measure of object tilt
angle (ϕ_1). The radius of
curvature of the fringes
(r_s) is a measure of object
bending radius (r_1), from
which the bending moment and
maximal object strain can be
calculated. Observe the
large information content of
Fig.5 compared to Fig.3

5. Maximal Strain by Fringe Rotation

The sandwich hologram was once more tilted so that the rigid
frame appeared fringe free. Then it was slightly rotated around
a vertical axis (ψ_2 of Fig.1). The rigid frame became covered
by vertical fringes while the fringes of the left bar got inclined to the right, the fringes of the right bar got inclined
to the left and the fringes of the middle bar bent into a "spoon
like" shape. This reconstruction (Fig.5) presents in one single
view much more information than can be found from Fig.3 which
represents the information content of a conventional double exposed hologram. The inclination of the fringes at each part of
the object is a measure of both sign and magnitude of object
tilt. The fringes are curved over those parts of the object where
the tilt angle varies. Thus the curvature of the fringes represents the derivative of the conventional fringe frequency and is

a measure of the amount of object bending. There exists a simple relation between fringe curvature and object bending moment from which object stress and strain can be evaluated [3].

6. Application

A two meter high milling machine with a weight of some two tons was recorded on a sandwich hologram. Between the two exposures a static load was applied to the machine. The fringes of Fig.6 reveal the resulting deformation in relation to a fixed reference frame. To the right of the machine is positioned a heavy piece of steel, which appears fringe free because it has not moved.

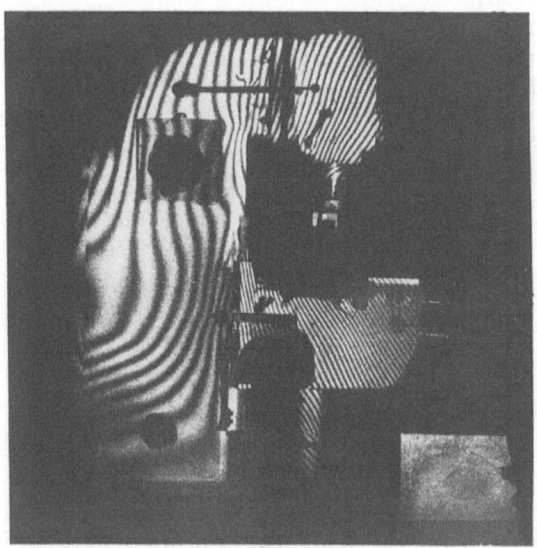

Fig.6. A milling machine has been loaded by a static force representing the cutting force. Every fringe represents a displacement of about 0.3 μm normal to the plane of the photograph. Straight lines represent a tilt around an axis parallel to the lines. Curved lines represent deformation. Displacements forwards or backwards can be distinguished directly

In Fig.7 the sandwich hologram has been tilted so that the deformation of the head of the machine can be studied without any influence from the large motions of the main body.

The fringes removed from the head now appear on the reference surface, the steelpiece at right. The fact that there are in Fig.7 more fringes on the table than in Fig.6 proves that the head and the table have tilted in opposite directions.

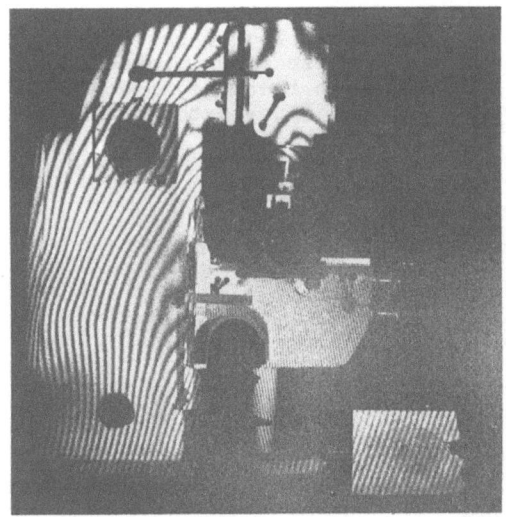

Fig.7. The sandwich hologram has been tilted so that the deformation of the machine head can be studied without any influence on the fringe pattern from the deformation of the total
machine

7. Contouring

Sandwich holography can be applied to any contouring method, but
we have concentrated our investigations to the method where the
point of illumination is displaced laterally between the two
exposures. This lateral shear was made by tilting a plane parallel glass plate that was introduced into the illumination beam
near its source. Each of the two hologram plates of the sandwich
represented one exposure and a tilt of the sandwich resulted in
a tilt of the interference fringe planes that intersected the
object. Figure 8 demonstrates the original fringe system with
no sandwich tilt and therefore the fringes are parallel to the
illumination direction. In Fig.8 the intersecting planes are perpendicular to the line of sight. In Fig.9 the sandwich has been
tilted until the interference fringe planes became parallel to
the top side of the cube respectively.

8. Theory and Calculations

The fringe motion caused by sandwich tilt can be explained by
the study of one single speckle ray which reacts to object movements as it was reflected by a small mirror fixed to its surface (Fig.1). Object bending will cause the speckle to be recorded at different positions on the two exposed hologram plates.
Any shearing or tilt of the sandwich that during reconstruction
repositions the two recordings of the speckle along the line of
sight towards the object eliminates the effects of the original

Fig.8. A small tilt of the sandwich hologram has rotated the intersecting interference planes so that they become perpendicular to the line of sight

Fig.9. The same reconstruction as that of Fig.8 but the intersecting planes are now parallel to the top side of the cube

speckle displacement and thus also eliminates the fringes indicating object motion. Thus fringes on the reconstructed object are depending on tilt direction, either subtracted from or added to, the original fringe pattern. The spacing and direction of the resulting fringes are analogous to a moiré effect of the original fringe pattern and the fringe pattern caused by sheared observation.

The following equations are limited to the study of displacements caused by object tilt that is parallel to the direction of illumination and observation. It has also been assumed that the sandwich tilt angle ϕ_2 is so small that $\sin\phi_2$ is equal to ϕ_2.

$$\phi_1 = \frac{d}{2Ln} \cdot \phi_2 \tag{1}$$

$$\phi_1 = \frac{d}{2Ln} \cdot \psi_2 \cdot tg\gamma \tag{2}$$

$$r_1 = \frac{2Ln}{d} \cdot \frac{I}{\psi_2} \cdot r_2 \tag{3}$$

$$\sigma = \frac{d}{2Ln} \cdot \frac{\psi 2Et}{2} \cdot \frac{1}{r_2} \tag{4}$$

The symbols are defined in Figs.1 and 4.

ϕ_1 = object tilt

ϕ_2 = sandwich tilt

d = thickness of hologram glass plate

L = distance from object to hologram plate

n = refractive index of hologram plate

ψ_2 = sandwich rotation around vertical axis

γ = fringe inclination to vertical

r_1 = object bending radius

r_2 = fringe radius of curvature

σ = object stress

E = modulus of elasticity

t = object thickness

9. Pulsed Holography

If the unique possibilities of double-pulsed hologram interferometry are to be fully utilized it is not enough to be able to freeze the motion during each exposure by using very short illumination pulses. Also needed is a method to eliminate the influence of the object's rigid-body velocity. Further on it is of great importance to be able to find the sign of the studied velocities. This sign is, in dynamics events, usually not known in advance in contrast to what is often the case with the displacement caused by static loading. Ordinary vibration patterns formed by timeaverage or by double-pulsed hologram interferometry do not reveal the phase relations between different vibrating surfaces. The reason is that the sign of motion (outwards or inwards) is missing in the conventional holographic process. To study the possibilities of solving these problems in a practical way we started applying the sandwich hologram method [1-3] to make vibration measurements on a hand-held drilling machine at work. To be able to study the dynamic behavior of such an unstable object we had to combine a double-pulsed ruby laser with a spinning hologram holder using the methods described in detail in [2] and [3] (Fig.10).

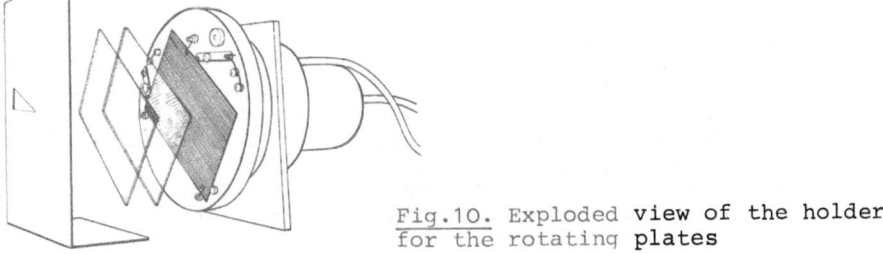

Fig.10. Exploded view of the holder for the rotating plates

Fig.11. Photo of the reconstructed image from a rotated sandwich hologram. The hand-held drilling machine was recorded during actual working conditions. The fringes are caused by a mixture of simultaneous rigid motions and local vibrations. The fringe-free reference surface indicates that in this position the sandwich hologram is identical to an ordinary double-exposed hologram

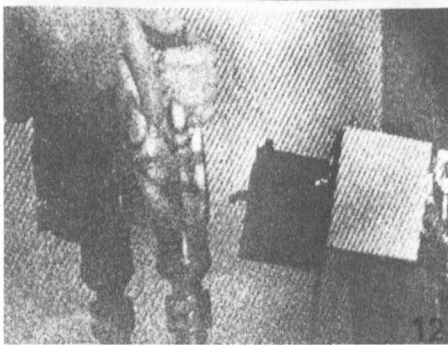

Fig.12. The sandwich hologram of Fig.11 has been tilted so the number of fringes at the studied area, around the tip of the index finger, is as low as possible. The eliminated fringes are seen at the reference surface and at the background. Apparently the rigid motion has been eliminated, so two vibrating areas with a node at the index finger are revealed

The photographs of Figs.11 and 12 are from reconstructions of one single sandwich hologram. Let us start by studying the photo of Fig.11.

The drilling machine is directed downwards and held by two hands, which are partly seen at the top left of the photo. The index finger of the right hand is almost vertical. A stable bright steel plate, which is fixed to the right of the machine, functions as a reference surface 1, 2, 3. Also, the background was stable enough to be used as a reference surface. The interference fringes on the hand-held machine are caused by a mixture of simultaneous rigid body motions and vibrations. In the following we will demonstrate how it is possible to eliminate the influence from the former and to concentrate the observation to local vibrations.

It is possible to transform the photo of Fig.11 into that of Fig.12 by tilting the sandwich hologram during reconstruction until the curved fringes close to the index finger break up into two circular fringe systems, indicating two points of maximal vibration amplitude separated by a nodal line at the index finger. The diagram of Fig.13 explains how the centers of the resulting ring systems represent the two points where the imaginary interference surfaces tangent the originally assumed flat object surface which has been curved by vibration.

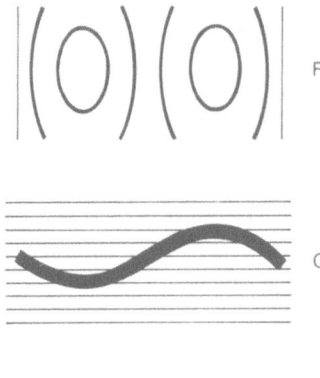

Fig.13. The pattern of Fig.3 is explained as caused by a flat object O vibrating in the form of an S and intersected by straight imaginary interference planes. The intersections cause the fringe pattern F as seen through the sandwich hologram H. The tangential points determine the centers of the two oval fringes. To simplify the picture the sandwich hologram is drawn without tilt when the object is also without tilt

In Fig.12 it is easy to see amplitude and mode of vibration, but that is almost impossible from Fig.11. By tilting the sandwich hologram the intersecting planes of Fig.13 are tilted and thus the centers of the oval fringes are displaced. In this way it is possible to find the phase relation between the vibrating areas.

10. Conclusions

The described possibilities of sandwich holography (Fig.1) make it especially interesting for biomedical investigations where the object cannot be immobilized in the same way as the metal objects usually studied by conventional holography. Therefore it is my belief that sandwich holography will greatly reduce the problems encountered in medical uses of holography.

In biomedical applications sandwich holography can be used to compensate for body motions so that a small local change can be studied in spite of large unwanted displacements. If Figures 2-5 represented the teeth of a human underjaw instead of metal bars, the tilt and displacement of every single tooth could be easily evaluated including its sign. This would be possible in spite of the relative motions of the patient's body, head and underjaw. The sign of motion is often unpredictable in biomedical investigations where the force might be unknown to the experimenter.

The tall machine studied at Figs.6 and 7 is of more than human size indicating motions of the total human body could be studied and measured. When a double-pulse laser is combined with the sandwich method of Figs.10-12 a body velocity of some 150 mm/sec can be compensated for, so that even then local changes as small as 0,3 μm and velocities smaller than 0,3 mm/sec can be studied. Using that same method also vibration modes induced e.g. by machines, sound or muscular activity could be holographed and the phase relations studied.

By contouring one can record images of three-dimensional objects with fringes that like the level lines of a map reveal the depth coordinates at any point of the object. Using such methods it is possible to record larger changes in shape for biomedical investigations, e.g. to study the growth or shrinkage of specimens. By using sandwich holography the imaginary intersecting planes can be tilted so that the optical sectioning of the object can be varied during the observation phase. This possibility should be of extra value in biomedical investigations where often the position of the object is difficult to predetermine.

References

1. N. Abramson: In *Optical Data Processing*, Topics in Applied Physics, Vol.23 (Springer, Berlin, Heidelberg, New York 1978) p.151
2. H. Bjelkhagen: Appl. Opt. *16*, 172 (1977)
3. N. Abramson, H. Bjelkhagen: Appl. Opt. *17*, 187 (1978)

Measurement of Vibration Waveforms Using Temporally Modulated Holography

C. Sieger and R. Röhler

Institut für medizinische Optik der Universität München
Theresienstr. 37, D-8000 München, Fed. Rep. of Germany

1. Introduction

The mechanical properties of the sound transmitting system, i.e. the ossicular chain and the fluid system of the inner ear, have an influence on the vibration behaviour of the tympanic membrane.

Any pathological alterations in this system may cause for example a change of the vibration amplitude of the membrane, which can be measured by the well known methods of holographic interferometry.

If these alterations occur, it could happen that the transmitting system does not obey any longer Hooke's law, that is, that deflection is no longer strictly proportional to force, an approximation, usually valid only for small motions. If such a nonlinear spring system is now excited by a sinusoidal force, it will no longer execute sinusoidal motions.

Also an examination of the vibration response of the tympanic membrane to excitation with sound containing higher harmonics could reveal information about the state of the system and the fidelity of its reaction.

Since these problems are also common and quite important in mechanical engineering, a lot of theoretical work has been done on the application of holography for the measurement of non-sinusoidal vibrations [1-5] . A method for the measurement of nonsinusoidal movements such as the variations in refractive index of water after local heating has been proposed [6]. In this method real time interferometry together with cinematography was used.Except a method for measuring small arbitrary displacements of specular reflecting objects [7] little work has been done on the problem of holographically measuring the vibration waveform itself.

DALLAS and LOHMANN [8] proposed a theoretical concept to calculate the temporal Fourier coefficients of a mechanical vibration from the temporal Fourier coefficients of the complex light amplitude emanating from the vibrating object.

In the next section of this paper we will show, how the theoretical concept of DALLAS and LOHMANN can be combined with the optical serrodyning technique, introduced by ALEKSOFF [9]. With this technique it is possible to measure arbitrary vibration waveforms also with higher harmonics and with larger amplitudes in the full human audio range.

In the third section we will present a method using strobe holography for measuring the Fourier coefficients of any vibration mode.

In the last section both techniques will be compared briefly.

2. Waveform Measurements Using $SSSC_n$ - Interferometry

We consider an arbitrary object motion $\Phi(x,t)$, periodic in time. In the usual time-average hologram interferometry this object motion is supposed to be a sinusoidal one. Since $\Phi(x,t)$ is periodic in time, it may be expanded in a temporal Fourier series

$$\Phi(x,t) = \sum_{m=-\infty}^{\infty} A_m(x)\exp(im\omega t)$$

In vibration holography, however, one measures the complex amplitude $O(x,t)$ of laser light, reflected by the vibrating object

$$O(x,t) = \exp(i\Phi(x,t))$$

This can also be expressed as a Fourier series

$$O(x,t) = \sum_{n=-\infty}^{\infty} O_n(x)\exp(in\omega t)$$

The problem of deducing the "mechanical" Fourier coefficients A_m from the "optical" coefficients O_n, which only can be measured, was solved by DALLAS and LOHMANN [8] with the following relationship

$$A_m = -\frac{i}{m}\sum_{n=-\infty}^{\infty} O_n O_{n-m}^* \tag{1}$$

This sophisticated algorithm relates the Fourier coefficients A_m of the object motion to the Fourier coefficients O_n of the complex light amplitude. If one knows the coefficients O_n of the complex amplitude $O(x,t)$ one is able to compute the temporal Fourier coefficients A_m of the mechanical vibration itself. The only problem is to measure the coefficients O_n.

This can be performed by means of multiplex holography, as CUTTER [10] did for very low frequencies (about 1 Hz). A moving grating, the velocity of which is synchronized with the fundamental vibration frequency acts as a multiple beam divider and as a frequency shifting device. Using this the intensities of the Fourier coefficients O_n of the object wave are recorded in the corresponding diffraction order n. Since the efficiency of the grating decreases with increasing diffraction order, also the reconstruction intensity and consequently the signal to noise ratio is considerably reduced at higher n values. Therefore only a few Fourier coefficients O_n (about n = 5) can be evaluated. Additionally this multiplex approach using a moving Ronchi ruling is hardly applicable to vibration frequencies in the audio range, since the minimum number of grating lines required will be greater than the product of the fundamental object vibration frequency and the exposure time.

ALEKSOFF [9] presented a method to translate the light frequency. This method is called optical serrodyning, since hereby the phase of the light is modulated in a sawtooth form (serra is the latin word for saw). If the phase is retarded or advanced by 2π and if the sawtooth frequency is f, then the light frequency is translated by $\mp\Delta f$, and the original frequency is completly suppressed. Therefore this technique is also called $SSSC_n$ - holography (Single-Sideband-Suppressed-Carrier). The index n labels the order of the sideband. Since the hologram acts as a narrow passband filter, any part of the objectspectrum, which originates from reflection at a vibrating object via the Doppler effect, can be filtered out by simply controlling the sawtooth frequency f. Hence all Fourier coefficients O_n of the spectrum can be measured by choosing the frequency in successive experiments as different multiples of the fundamental frequency. Since - in principle - there exists no theoretical limitation of the amount of the frequency translation, the higher coefficients can be measured, too.

This is shown by a record (Fig.1) of the 14-th sideband of the objectspectrum, produced by a small membrane ($\emptyset$ = 10 mm) vibrating with 1050 Hz.

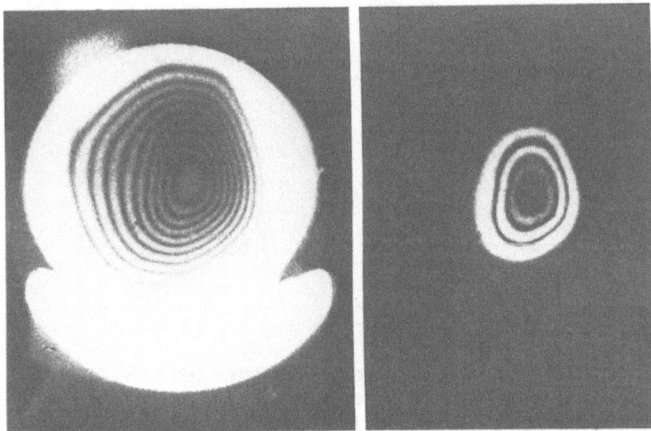

Fig.1 On the left a time-average hologram, i.e. a $SSSC_0$-hologram, and on the right the corresponding $SSSC_{14}$- hologram

From Fig.1 once more one can see, as already demonstrated [9], that the SSSC-holography is useful for the measurement of large amplitude as well as of small amplitude vibrations, if one takes a first order sideband hologram (Fig.2).

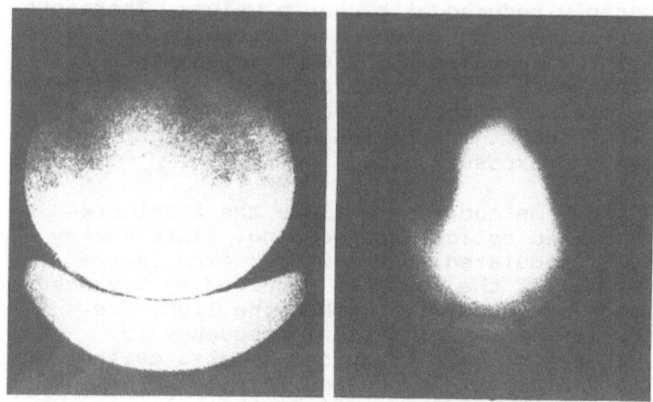

<u>Fig.2</u>

<u>Fig.3</u>

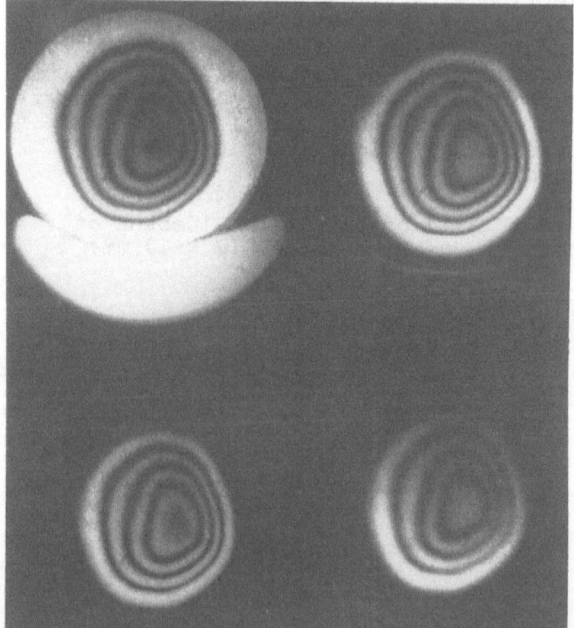

<u>Fig.2</u> Detecting small-vibration amplitudes. On the left with a time-average record, on the right with a $SSSC_1$- hologram of the membrane, vibrating with the same amplitude (about 20 nm)

<u>Fig.3</u> 0-th, 1-st, 2-nd and 3-rd order sideband of a sinusoidally vibrating membrane

Unfortunately with any recording process one stores only the intensities of the various O_n instead of the complex amplitudes, which have to be known for calculating the Fourier coefficients of the vibration. Since the phase measurement is rather cumbersome, we will briefly consider, which information one can get from the intensity measurement of the coefficients O_n alone. One can calculate the coefficients for a given waveform. For a sinusoidal vibration the intensities of the coefficients of order n are for example proportional to the square of the n-th order Bessel functions. One can now holographically record some coefficients and can then compare their microdensitometer traces with the theoretically predicted intensities (Fig.3 and 4). Deviations from the sinusoidal waveform cause differences between the calculated and the measured intensities. The experiments show, however, that this method is not very sensitive, at least not for detecting small deviations from the pure sinusoidal vibration.

Useful on the other hand are the intensity measurements of corresponding sidebands with opposite signs for the detection of unsymmetrical vibrations, i.e. vibrations, where the movement in one direction does not equal the return movement (Fig.5). An instructive example of such a vibration is the sawtooth producing only one sideband, whereas the intensity of the corresponding sideband with opposite sign is zero.

For measuring the phase angle Δ, which together with the intesity measurement is necessary to retrieve the complex amplitude, there exist 2 methods (see Fig.6).

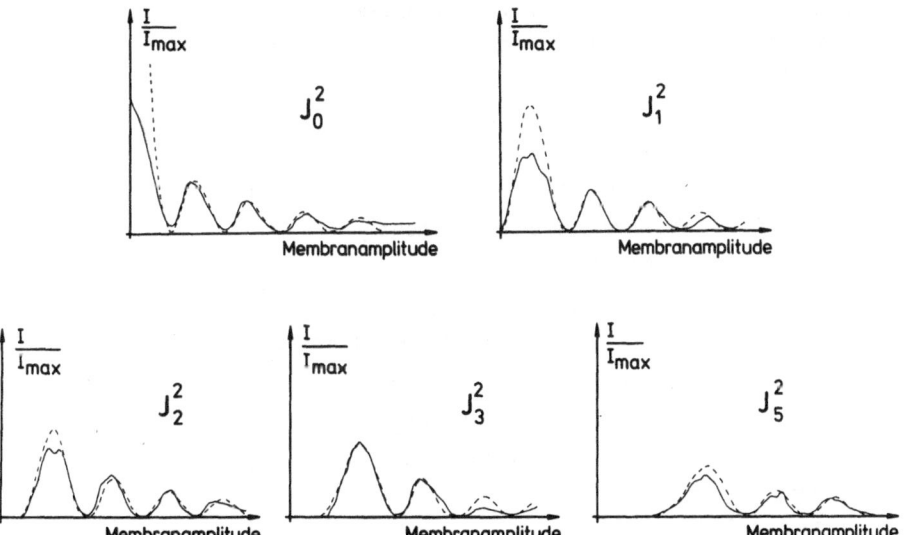

Fig.4 Measured (full line) and theoretically predicted (dotted line) intensities of several coefficients of the objectspectrum, produced by a sinusoidally vibrating membrane

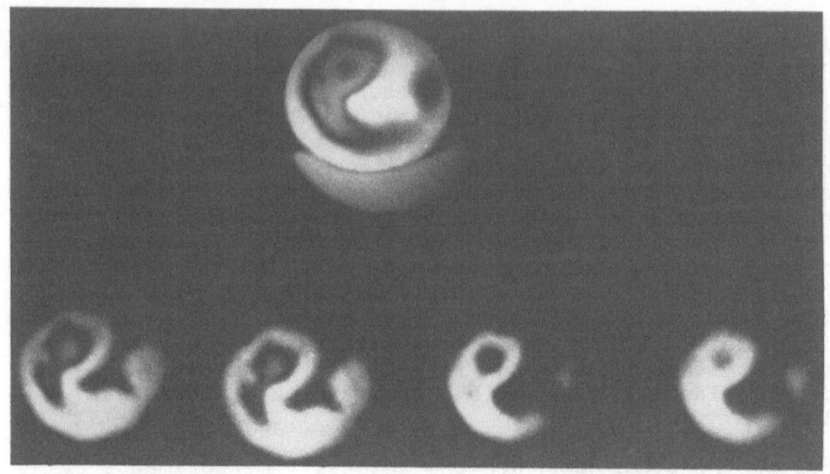

<u>Fig.5</u> An example of an unsymmetrically vibrating membrane.
Corresponding sidebands (lower row) do not equal one another

1. 4 Intensity Measurements in Real-Time
   (see D.Cutter, Proc. of the eng. uses of coherent
   optics. Glasgow 8.-11. April 1975)

   or Using Holographic Addition

2. Direct Phase Measurement in Real-Time, Using
   a Variable Phase Shifting Device
   (e.g. a Piezo-Electric Element)

$$I_1 = O_n^2 \qquad O_n : \text{Amplitude of the n-th Side-Band of the Membrane}$$

$$I_2 = O_o^2 \qquad O_o : \text{Amplitude of the Stationary Membrane}$$

$$I_3 = |O_o + O_n\, e^{i\Delta_n}|^2 \qquad \Delta_n : \text{Phase Difference Between the Two Waves}$$

$$I_4 = |O_o\, e^{i\frac{\pi}{2}} + O_n\, e^{i\Delta_n}|^2 \qquad \text{Sum of the Two Waves After Shifting the Phase by } \pi/2$$

$$\tan \Delta = \frac{I_4 - I_1 - I_2}{I_3 - I_1 - I_2}$$

<u>Fig.6</u> Two possibilities of phase measurement

The first method is to perform four intensity measurements I_1 to I_4. I_1 is the intensity of the n-th sideband, I_2 the intensity of the object at rest. The intensities I_3 and I_4 were measured in real time or by means of a double-exposure (holographic addition) and represent the superposition of the object at rest and the n-th sideband. For the measurement of I_4 an additional phase shift of $\pi/2$ is introduced. From these intensity measurements the phase angle can be calculated (Fig.6).

For the second method, conceptionally more direct, one needs a second phase shifting device, e.g. a piezoelectric element with a small mirror attached to it. The applied voltage necessary to get an intensity minimum in the real time reconstruction between the stationary object and the n-th temporal sideband is a direct measure of the phase difference Δ.

With the knowledge of the Fourier coefficients of the complex light amplitude one can easily calculate, preferably by means of a computer, the temporal Fourier coefficients of the waveform itself.

3. Waveform Measurement Using Stroboscopic Holographic Interferometry

This method for waveform measurement uses the strobe holography [11]. 12 double-pulse interferograms have to be taken using a pulse laser, a Pockels cell together with a cw-laser or the method proposed by FRYER [12]. The first pulse M_1 must always be performed at the same but arbitrary point of the vibration cycle. Then the corresponding second pulses M_n have to run through the entire vibration cycle released at 12 equidistant time intervalls. From these interferograms the instantaneous amplitudes can be determined by means of (2) or (3).

$$I = I_B \cos^2 \frac{C(M_1-M_2)}{2} = I_B \frac{1}{2}\left\{1 + \cos(C(M_1-M_2))\right\} \qquad (2)$$

$$I = I_B \sin^2 \frac{C(M_1-M_2)}{2} = I_B \frac{1}{2}\left\{1 - \cos(C(M_1-M_2))\right\} \qquad (3)$$

where I describes the intensity of a point (x,y) on the reconstructed image, I_B is the image intensity of the same point on the stationary object, M_1 and M_2 describe the 2 fixed positions, occupied by the membrane during the stroboscopic light flashes. C is a constant, given by

$$C = (2\pi/\lambda)(\cos\varphi_1 + \cos\varphi_2) \qquad (4)$$

where λ is the laser wavelenght, φ_1 and φ_2 are the angles between the vibration vector of the membrane and the directions of illumination and observation, respectively.

Calculation Scheme

$M - M_n = \Delta M_n \qquad n = 12$

ΔM_0	ΔM_1	ΔM_2	ΔM_3	ΔM_4	ΔM_5	ΔM_6	S_0 S_1 S_2 S_3	d_1 d_2 d_3
ΔM_{11}	ΔM_{10}	ΔM_9	ΔM_8	ΔM_7			S_6 S_5 S_4	d_5 d_4
Sums	S_0 S_1 S_2 S_3 S_4 S_5 S_6						σ_0 σ_1 σ_2 σ_3	ε_1 ε_2 ε_3
Differences	d_1 d_2 d_3 d_4 d_5						δ_0 δ_1 δ_2	γ_1 γ_2

			Cosine			Sine			
1	σ_0 σ_1 σ_2 σ_3	δ_0	$\sigma_0 - \sigma_3$ δ_0 δ_2	ε_3			ε_1 ε_3		
0.866		δ_1				ε_2	γ_1 γ_2		
0.5		δ_2	$-\sigma_2$ σ_1		ε_1				
Sums	I II	I II	I II	I II	I II	I II	I II	I II	
I + II	$6A_0$	$6A_1$	$6A_2$		$6B_1$	$6B_2$			
I – II	$12A_6$	$6A_5$	$6A_4$	$6A_3$	$6B_5$	$6B_4$	$6B_3$		

$$\Phi(x,t) = \frac{A_0}{2} + \sum_{m=1}^{6} A_m \cos m\omega t + \sum_{m=1}^{5} B_m \sin m\omega t$$

Fig.7 Evaluation scheme for the Fourier analysis of waveforms

Eq.(2) is valid for the usual double-pulse or strobe holography, whereas (3) describes the intensity of the reconstructed
image, when additionally the holographic subtraction technique
is applied. For this a phase change of π is introduced to either
the reference or object wave between the two corresponding
series of light pulses. This produces a dark background of
stationary regions, whereas slightly vibrating regions appear
bright according to (3). This subtraction method is useful for
the measurement of small-amplitude vibrations. The 12 values M_n
of the instantaneous amplitudes, measured as described above,
were inserted in an evaluation scheme (Fig.7) yielding at once
5 Fourier coefficients of the mechanical waveform.

As demonstrated by VON BALLY [13] it is possible to record in
vivo double-pulse interferograms of the human tympanic membrane
with good quality. By means of the method described in this
section, vibration waveforms of normal and pathological tympanic
membranes can be measured.

4. Comparison of Both Methods

Fig.8 shows the used set-up for waveform measurements. It is a
very common one, only containing additionally a Pockels cell for
stroboscopic object illumination (necessary for method 2), a

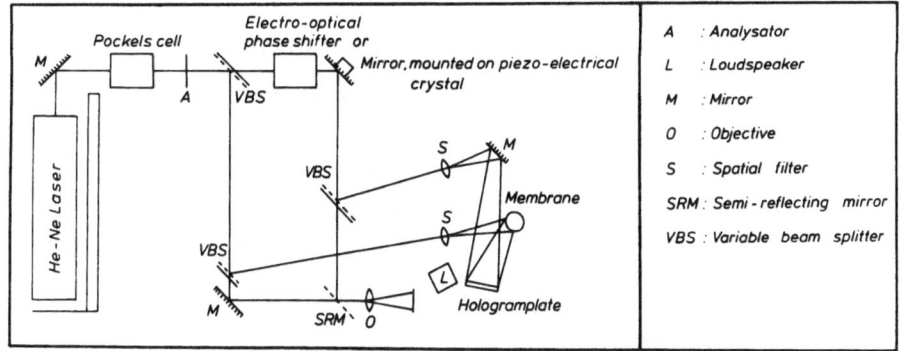

A	: Analysator
L	: Loudspeaker
M	: Mirror
O	: Objective
S	: Spatial filter
SRM	: Semi-reflecting mirror
VBS	: Variable beam splitter

Fig.8 A holographic system for waveform measurements

piezoelectric element (for 1 and 2) and an electro-optical phase shifter (method 1). To control the amount of the phase shift, the set-up includes a Mach-Zehnder interferometer. It is favourable to lead the object and reference beam close together, thereby reducing the undesired effects by air drifts.

The first method has the serious drawback, that only small vibra- tions can be measured with reasonable expenditure. For the measurement of a vibration with peak to peak amplitude of one wavelength, one has to measure at least the coefficients O_n with $|n| \leq 13$ to measure the amplitude within 2% [10]. This means, how- ever that one has to carry out 27 intensity and 26 phase measure- ments. Even if all these measurements were performed in real time with only one exposed hologram (we will demonstrate this in a subsequent paper), this is much more troublesome than to perform 12 double-pulse interferograms, independent of the amplitude. Furthermore, the first method is not applicable to in vivo measurements.

An advantage of the first method is that it makes feasible also non-harmonic vibration analysis. Additionally, as theore- tically demonstrated by [8] and [10], also a direct optical display of object motion parameters on the reconstructed image is possible.

The accuracy of measuring the waveforms by the strobe method is essentially limited by uncertainties in the identification of the fringe order, which can be interpolated in our holograms for any point of the membrane down to 1/6 of a fringe. That means that the amplitude can be measured with an accuracy of ± 26 nm, independent of the amplitude. By applying the dual-frequency method [14] this error can be reduced at least by the factor 15.

The first method is much more inaccurate, since the phase measurements are affected with a r.m.s. deviation of about $15°$, which can be improved by repeated measurements. Furthermore intensities were measured instead of amplitudes, a fact that requires a strictly linear recording process. Furthermore this

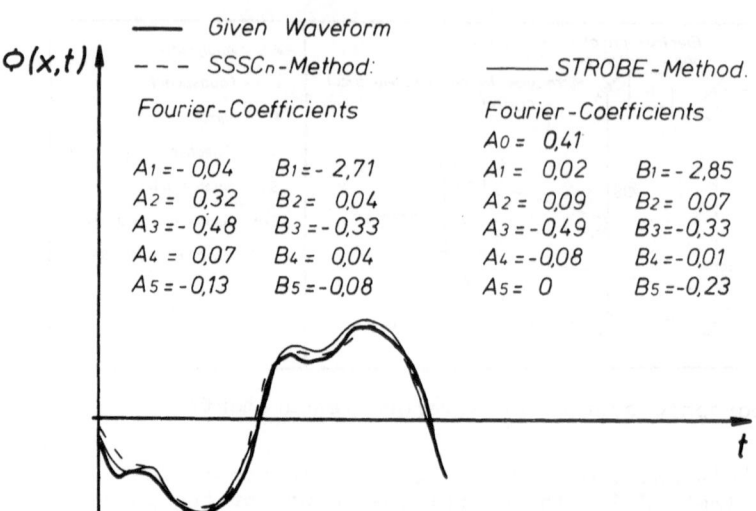

Fig.9 Measurement of a given waveform

method is only applicable to small vibrations. All these items contribute to the considerably high error inherent in this method.

Finally, we will give two examples of waveform measurements. First a loudspeaker was driven with a known signal and then measured by means of the two methods (Fig.9).

Then an unknown waveform arising in a one-side damped membrane [15], which is excited acoustically with a sine-signal, was examined.

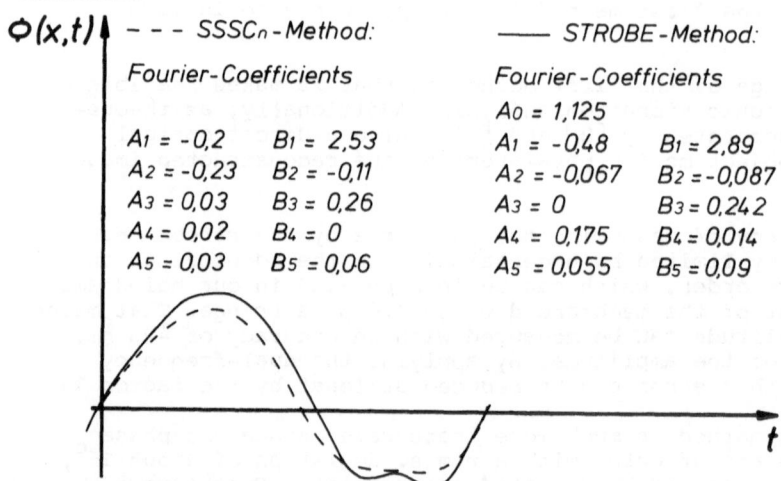

Fig.10 Vibration waveform of a membrane, vibrating unsymmetrically to its position at rest

By means of the two methods outlined, not only amplitudes, but also oscillation waveforms can be measured. The vibration response, however, of a sound generating or sound transmitting system to a certain signal is a criterion for its quality. Therefore the examination of the vibration waveforms of the tympanic membrane could become a sensitive tool for detecting disorders in the sound transmitting system of the middle ear.

5. References

1 K.A. Stetson, J.Opt.Soc.Am. 61 (1971) 1359
2 K.A. Stetson, J.Opt.Soc.Am. 62 (1972) 297
3 A.D. Wilson , J.Opt.Soc.Am. 61 (1971) 924
4 P.C. Gupta and K. Singh, Appl. Opt. 14 (1975) 129
5 J. Janta and M. Miler, Optik 36 (1972) 185
6 G. Pierattini, Opt. Commun. 5 (1972) 41
7 Y. Ohtsuka and I. Sasaki, Opt. Commun. 10 (1974) 362
8 W.J. Dallas and A.W. Lohmann, Opt. Commun. 13 (1975) 134
9 C.C. Aleksoff, Appl. Opt. 10 (1971) 1329
10 D. Cutter, Thesis of the Friedrich-Alexander-Universität
            Erlangen – Nürnberg (1976)
11 E. Archbold and E.A. Ennos, Nature 217 (1968) 942
12 P.A. Fryer, Appl. Opt. 11 (1972) 1642
13 G. von Bally, this issue
14 R. Dändliker, B. Ineichen and F.M. Mottier, Opt. Commun. 9
            (1973) 412
15 R. Röhler and C. Sieger, Opt. Commun. 25 (1978) 297

Compensation for Rigid Body Motions in Holographic Interferometry

G. Ferrano, G. Häusler

Physikalisches Institut der Universität Erlangen
Erwin-Rommel-Str. 1, D-8520 Erlangen, Fed. Rep. of Germany

1. Introduction

In this paper we report about holographic interferometry in the
presence of large rigid body motions.
A medical question was asked by the Institute of Orthopaedics
in Erlangen, concerning the change of deformation of a human
bone, that is caused by the implantation of a hip-joint
prosthesis. This problem is difficult in that if we want to
measure the deformation under real circumstances, such as with
great forces and unprepared bones, we have problems with large
rigid body motions [1].

To attack this problem, we used the set-up shown in fig.1.

The laser beam is divided by a beam splitter. Because of the
lens L the object is illuminated by a plane wave. The
reference wave is also planar. The two mirrors in the
illumination and in the reference beam serve to compensate for
tilts and lateral translations. In order to control the
compensations, we work in real-time.

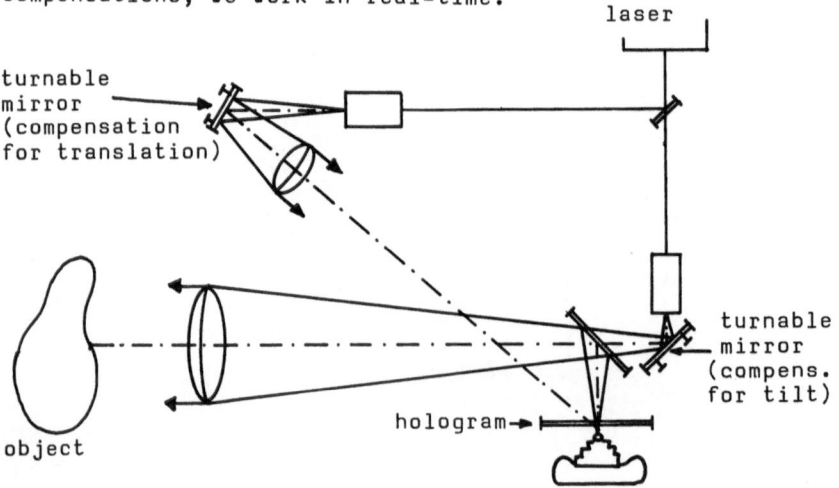

Fig.1 Set-up for holographic interferometry, including
posibilities of compensation.

2. Compensations for Rigid Body Motions

2.1. Tilt Compensation

Because of the geometry of our set-up ,tilts produce parallel
fringes, which only increase the total number of fringes
without adding any information.Since we illuminate our object
with a plane wave, a tilt of the object can be compensated for
by tilting the illumination wave. This is done by a turnable
mirror. The effect is shown in the next three pictures.

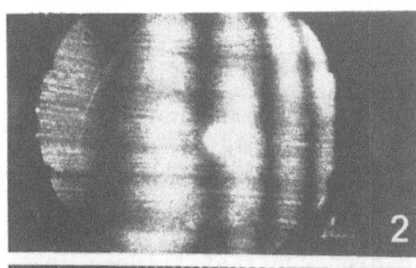

Fig.2 Interference
pattern

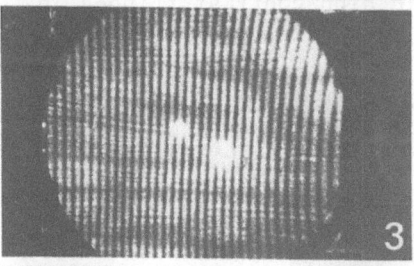

Fig.3 Interference
pattern, including
tilt

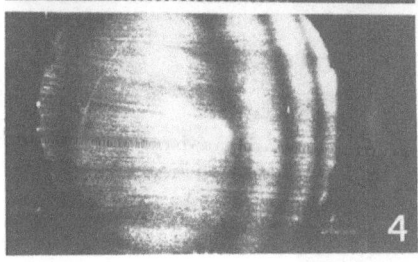

Fig.4 Interference
pattern, after tilt
compensation

This compensation works for large objects and tilts nearly
without fault. We were able to compensate for a tilt of 1.8
degrees over an object range of 200mm. This tilt would
normally have produced about 25000 fringes [2,3].

2.2. Compensation for Lateral Translations

A lateral translation produces a much different kind of effect.
The interference fringes move away from the object. Because we
have to focus exactly on the object, the contrast of the

fringes therefore decreases. We can avoid this by manipulating the reference wave in such a way, that the reconstructed holographic image has the same lateral translation as the object. The question is, how to manipulate the reference wave. In order to find the best method we calculated the "lens-equation" for holography. The result was, that we have to tilt the reference wave in order to produce a lateral translation for the object. This is done by a mirror in the reference beam [3].

In doing so, we unfortunately produce image aberrations, distortion and curvature of field.

2.2.1. Distortion in Holographic Reconstruction

This distortion is perpendicular to our sensitivity vector. The result is, that the object and the hologram reconstruction coincide only in a limited area. This area depends on the compensated translation, the distance between the object and the hologram and on the dimension of the object itself. The effect is shown in the next figure. Here the compensated translation was 1mm and the distance between the object and the hologram 400mm. We were able to obtain high contrast fringes in an area of about 100mm in diameter.

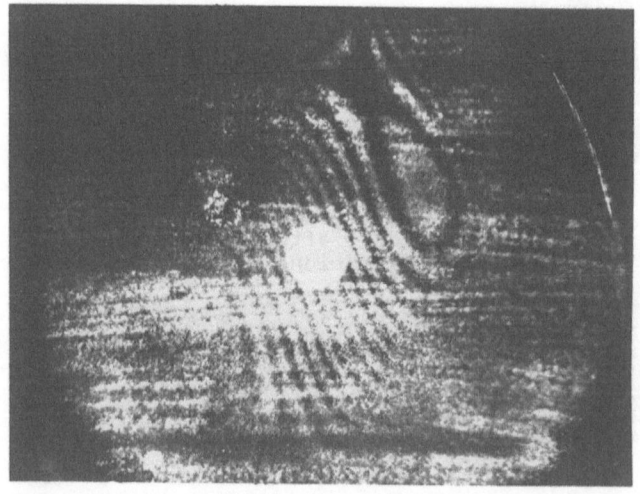

Fig.5 Effect of distortion
Because of distortion, high contrast fringes exist only in a certain area.

2.2.2. Curvature of Field

The curvature of field is the aberration with greater importance, because it produces changes parallel to our sensitivity vector. It falsifies the interferogram in that the aberrated output can be interpreted as an object deformation.

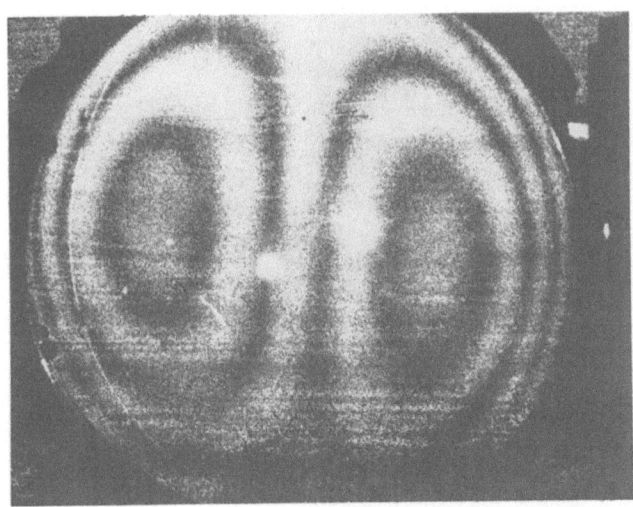

<u>Fig.6</u> Apparent object deformation, caused by compensation for translation (200μ)

With an optimal geometry, these faults can be reduced to obtain the desired measurement accuracy.It depends, of course, on the range of translations which can occur. One can calculate the region where the errors are low enough that one can speak of "faultless" measurement. In our case it is a region of 80mmx80mmx80mm (with fault less than 0.1μ).

This is shown in the next two figures. A plane plate is tilted to produce the interferogram shown in fig.7. The object is then shifted by 200μ to simulate a rigid body motion. Compensation for the translation produces an aberrated interferogram as in fig.8. Within the circled region the aberrations are not severe and the compensated interferogram looks like the unshifted original one.

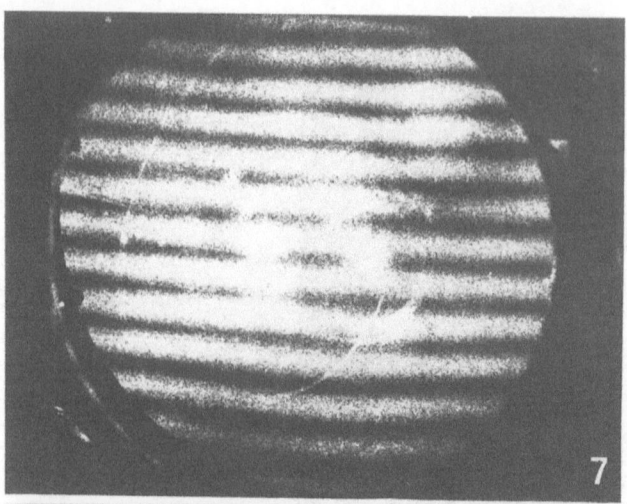

Fig.7

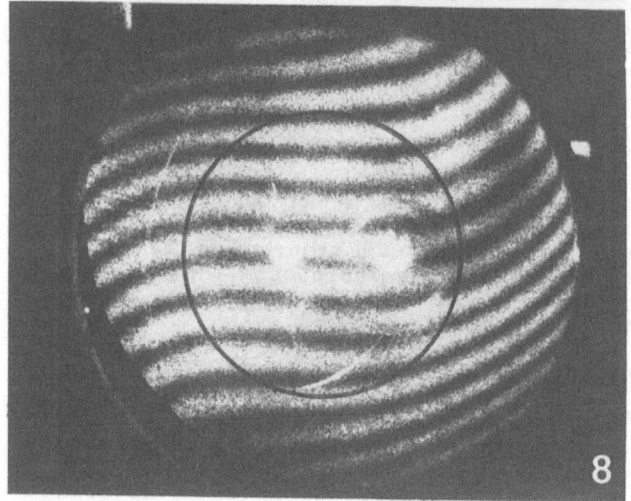

Fig.8

Fig.7 Interferogram of a tilt

Fig.8 Interferogram, after 200μ compensated translation

References

[1] G. Häusler, T. Schwenk, K. Seidel, Holografische Deforma-
tionsmessungen zur Optimierung von Hüftgelenkimplantaten.
Teratologische Forschung und Rehabilitation Mehrfachbehin-
derter, Symposium Münster (1976) 349

[2] L. Kersch, Advanced Concepts of Holographic Nondestructive
Testing, Mater. Evaluat. XXIX (1971) 125

N. Abramson, Sandwich Hologram Interferometry, a new Dimension in Holographic Comparison.
Applied Optics 13 (1974) 2019

K. Piwernetz, Untersuchungen über das Deformationsverhalten
menschlicher Wirbelkörper bei statischer Belastung mittels
holografischer Interferometrie
Teratologische Forschung und Rehabilitation Mehrfachbehin-
derter, Symposium Münster (1976) 321

[3] G. Ferrano, G. Häusler, Kompensation von Ganzkörperbewegun-
gen bei der Holografischen Interferometrie
Optik (accepted for publication)

Automatic Iconics Measuring System Applied to Biostereometrics

P. Meyrueis, C. Liégeois, and M. Grosmann

Groupe de Recherche et d'Expérimentation en Photonique Appliquée,
Laboratoire de Spectroscopie et d'Optique du Corps Solide
(associé au C.N.R.S. no 232)
Université Louis Pasteur
3, rue de l'Université, F-67084 Strasbourg Cedex, France

M. Lasalle

Société Matra
93, avenue Victor Hugo, F-92503 Rueil Malmaison, France

1. Working Principle

When three-dimensional measurements of an object are to be made,
two techniques are commonly used :

a) In mechanics the object itself is usually available
throughout the duration of the measurements, a mechanical
finger, which is moved over the object, can be used. The
measurements are made by recording the coordinates given
by the finger as it is displaced. These measurements thus
require a physical contact with the object. This contact
may be destructive.

b) In case where no contact with the object can be tolerated,
and where the object itself cannot be immobilized,
techniques of short-distance photogrammetry can be used.
However, this method also presents inconveniences :

. certain object forms are not easily photographed
from two different angles.

. it is difficult to produce stereoscopic plates of
objects moving at high translation velocities in
normal lighting.

. phase phenomena cannot be photographed.

Because we are actively involved in the field of research on
application of holography, we have naturally become interested
in the possibilities offered by holography in the field of
three-dimensional measurement.

Holography is, in effect, a technique permitting the
recording and reconstruction, in simple fashion, of the image
of a three-dimensional object. If certain precautions are taken,
a stigmatic image, life-sized, is reconstructed, and we thus
obtain, by means of a very simple setup, the equivalent of the
"stereomodel" given by the conventional processes of photogrammetry.

The exploitation of the image can be made by means of the
two distinct techniques which have been studied.

2. Holographic Reconstructor working on the Virtual Image
Fig. 1.

This technique remains very similar to that used in photogrammetry. It consists in visually bringing a spatial marker into
contact with the point of the holographic image whose spatial
coordinates we want to know.The observer looks at the reconstructed image through the holographic film. The luminous
spatial marker is the end of an optical fiber 250 μ" in diameter,
mobile along 3 orthogonal axes.

This displacements in space of the end of the fiber are
simultaneous displayed on 3 counters and the coordinates placed
in a computer memory. The displacements along 3 axes are
controlled by means of two manual commands.

The maximum displacements which can be obtained with our
prototypes are 230 mm along OX, 300 mm along OY, 230 mm along
OZ.

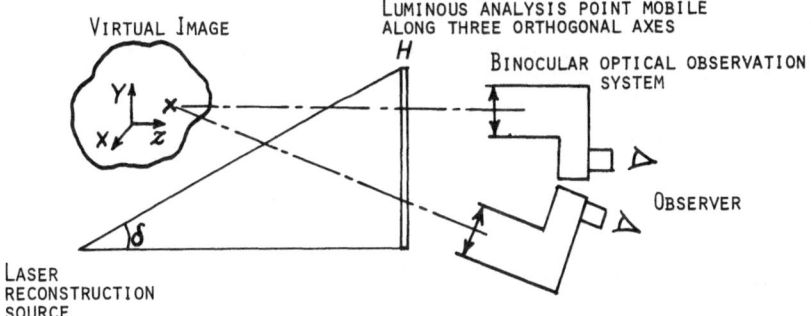

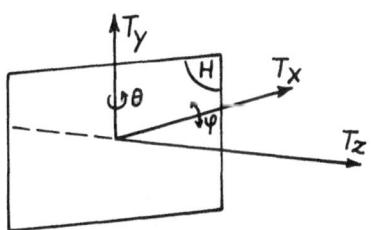

Fig.1. Principle of hologrammetric reconstructor in the virtual image

2.1 Results obtained

We first tried to evaluate the precision in repeatedly aiming
at a single point, by studying the dispersion of the results
obtained.

The observation of the correspondence was made either with
the naked eye or by means of a binocular optical system (8x)
with pupils of 20 mm.

The dispersions obtained from a statistical series of aimings
with the 8x optical system and with all adjustments destroyed
after each measurement are :

axis	global dispersion
X	± 0.02 mm
Y	± 0.02 mm
Z	± 0.10 mm

We also studied the precision with which the holographic
film must be repositioned in this mounting.

$$\Delta\theta = \pm 70' \text{ (rotation around the TY axis)}$$

$$\Delta\psi = \pm 30' \text{ (rotation around the TX axis)}$$

translation along the TX and TY axes = ± 2 mm

A translation along TZ axis modifies the enlargement between
the object and its holographic image. This translation along
TZ must be limited to ± 10 mm to obtain sufficient quality
of resolution, which means that the reconstructor must necessarily function at an enlargement of 1X.

Trials were also made working at reconstruction with wavelengths different from those used in recording.

3. Holographic Reconstructor working on the Real Image
Fig. 2

It is possible, in taking certain precautions in recording,
to reconstruct a real holographic image. In this case, we have
direct access to the image volume. We can consider using techniques of aiming identical to those used in metrology (in
coherent light), or we can place the photocathode of a vidicon
tube directly in the reconstructed image. The aimings are then
made on the screen of the television monitor.

Theoretically the precision of the longitudinal aimings is
greater if a conjugation lens is used between the hologram and
the enlarging tube. Unfortunately, the speckle phenomena
quickly limit the enlargements which can be used. Efforts to
reduce the bothersome effect of the speckles by optical or electronic processes have given satisfactory results only with
expensive apparatus.

To increase the precision of the longitudinal aimings, we
were led to develop a stereoscopic viewfinder. This viewfinder
is mounted on a coordinatograph permitting its displacement
along 3 orthogonal axes and controlled by the operator by means
of the same system used in the deplacement of the optical
fiber.

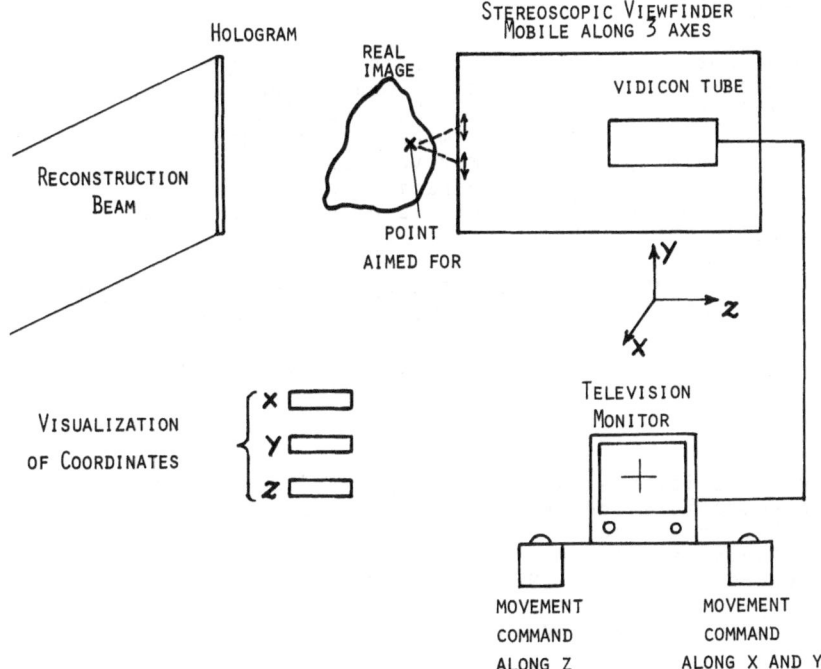

Fig.2. Principle of hologrammetric reconstructor in the real image

3.1 Results obtained

a) *Precision of aimings*

We measured the displacement obtained over statistical series of aimings along :

axis	maximum displacement observed
OX	0.02 mm
OY	0.02 mm
OZ	0.05 mm

b) *Precision of measuring of distance in space*

There is an error of about 1/100 between the size of the real object and its holographic image. The error is systematic and can be corrected by recording a standard of known size at the same time as the object. The precision of measurement over a distance is thus twice as bad as on a point.

c) *Precision of repositioning of the hologram*

Since the reference beam is parallel,the precision of repositioning of the hologram in its plane is unimportant. In rotation, the precisions of repositioning are in the order of 20 minutes of arc.

4. Conclusions

The experiments performed show the possibilities offered by holography enabling us to make precise measurements in space by means of fairly simple apparatus. We have also seen the advantages of working on the real image : on one hand, the precision of the aimings can be improved; on the other, employing a video system in the operation permits us to semi-automatically follow repetitive details existing or to be created optically on the object.

Setups for measurement on holographic images should be considered as complementary to photogrammetric instruments.

We must thank the D.R.M.E., which helped finance these studies and authorized us to publish the results.

References

J.P. Agnard, A. Boisvin et A.J. Brandeberger, "Obtention de pointes stéréoscopiques de précision sur l'image holographique virtuelle" "Hologramétrie".
Compte rendu Symposium International "application de l'Holographie", Besançon, Juillet 1970.

M. Françon, "Holographie", ed. Masson (1969).

P. Smigielski, C. Vienot, H. Royer, "Holographie optique", ed. Dunod (1971).

P. Meyrueis, Thèse, 1976, U.L.P., Strasbourg.

Index of Contributors

H. Haken
Synergetics

An Introduction

Nonequilibrium Phase Transitions and Self-Organization in Physics, Chemistry and Biology
Springer Series in Synergetics

2nd enlarged edition. 1978. 152 figures, 4 tables.
XII, 355 pages
ISBN 3-540-08866-0

"...In this book an introduction is given to the basic physical ideas and mathematical methods to be used. The Text is imaginatively written and well illustrated by an amazing variety of examples drawn from such diverse fields as laser physics, fluid dynamics, mechanical engineering, chemical reactions, ecology and morphogenesis. In particular I appreciated the care that Professor Haken had taken in introducing and making accessible topics and examples from other disciplines so that no specialized knowledge was required to read them. Correspondingly the mathematical tools have also been introduced with similar care and taken by themselves would form a good introduction to the study of probability and information theory, stochastic equations, Fokker Planck and master equations. There is also some discussion of the phase plane analysis of dynamic processes together with an elementary account of catastrophe theory. The formation of organized structures out of chaos represents fascinating and challenging problems. Professor Haken is to be congratulated in producing such a readable introduction to a subject still in its infancy."

Physics Bulletin

Synergetics

Far from Equilibrium

Proceedings of the Conference Far from Equilibrium: Instabilities and Structures, Bordeaux, France, September 27–29, 1978

Editors: A. Pacault, C. Vidal
Springer Series in Synergetics.
With contributions by numerous experts

1979. 109 figures, 3 tables. IX, 175 pages
ISBN 3-540-09304-4

The tutorial lectures held at this conference introduce newcomers to this new field of research, which brings together thermodynamics of irreversible processes, the theory of phase transitions, bifurcation analysis and catastrophe theory. Research papers provide a deeper insight into both experimental and theoretical problems and results. The examples given are related mainly to hydrodynamics and chemistry; nevertheless, the analogies with other disciplines are quite obvious and already well-known.

Synergetics

A Workshop

Proceedings of the International Workshop on Synergetics at Schloß Elmau, May 2–7, 1977

Editor: H. Haken
Springer Series in Synergetics.
With contributions by numerous experts

1977. 136 figures. VIII, 274 pages
ISBN 3-540-08483-5

Synergetics: A Workshop presents the latest theoretical and practical advances in synergetics – a relatively new field of interdisciplinary research which studies the self organized behavior of systems leading to the formation of structures and functionings. The authors of each of the contributions represent their most recent experimental and theoretical results in a way which can easily be understood by scientists working in different disciplines. The many additional references will enable research workers to quickly penetrate facts and methods dealing with pattern formation and self-organization. Particular emphasis is given to the profound analogies among phenomena of different disciplines, such as physics, astrophysics, chemistry, biology, and sociology.

Structural Stability in Physics

Proceedings on Two International Symposia on Applications of Catastrophe Theory and Topological Concepts in Physics, Tübingen, Fed. Rep. of Germany, May 2–6 and December 11–14, 1978

Editors: W. Güttinger, H. Eikemeier

1979. 108 figures, 8 tables. VIII, 311 pages
ISBN 3-540-09463-6

These contributions discuss recent applications to physical systems of topological concepts derived from structural stability. Catastrophe and singularity theory play a central role. The diverse physical systems considered exhibit analogous behavior on various scales. Owing to the broad scope of this rapidly expanding field, the papers cover a wide spectrum: topological aspects of wave motion including fractal regimes and optical diffraction catastrophes, catastrophe theory in infinite-dimensional systems and semiclassical quantum theory, stable defects and dislocations, structural stability in statistical mechanics, catastrophe theory and phase transitions, solitons in physics, chaos, and applications of catastrophe theory in biophysics and pattern recognition.

Springer-Verlag
Berlin Heidelberg New York

Digital Picture Analysis

Editor: A. Rosenfeld
1976. 114 figures, 47 tables. XIII, 351 pages
(Topics in Applied Physics, Volume 11)
ISBN 3-540-07579-8

Contents:
A. Rosenfeld: Introduction. – *R. M. Haralick:* Automatic Remote Sensor Image Processing. – *C. A. A. Harlow, S. J. Dwyer III, G. Lodwick:* On Radiographic Image Analysis. – *R. L. McIlwain, Jr.:* Image Processing in High Energy Physics. – *K. Preston, Jr.:* Digital Picture Analysis in Cytology. – *J. R. Ullmann:* Picture Analysis in Character Recognition.

Optical Data Processing

Applications
Editor: D. Casasent
1978. 170 figures, 2 tables. XIII, 286 pages
(Topics in Applied Physics, Volume 23)
ISBN 3-540-08453-3

Contents:
D. Casasent, H. J. Caulfield: Basic Concepts. – *B. J. Thompson:* Optical Transforms and Coherent Processing Systems – With Insights from Crystallography. – *P. S. Considine, R. A. Gonsalves:* Optical Image Enhancement and Image Restoration. – *E. N. Leith:* Synthetic Aperture Radar. – *N. Balasubramanian:* Optical Processing in Photogrammetry. – *N. Abramson:* Nondestructive Testing and Metrology. – *H. J. Caulfield:* Biomedical Applications of Coherent Optics. – *D. Casasent:* Optical Signal Processing.

T. Pavlidis

Structural Pattern Recognition

1977. 173 figures, 13 tables. XII, 302 pages
(Springer Series in Elektrophysics, Volume 1)
ISBN 3-540-08463-0

Contents:
Mathematical Techniques for Curve Fitting. – Graphs and Grids. – Fundamentals of Picture Segmentation. – Advanced Segmentation Techniques. – Scene Analysis. – Analytical Description of Region Boundaries. – Syntactic Analysis of Region Boundaries and Other Curves. – Shape Description by Region Analysis. – Classification, Description and Syntactic Analysis.

X-Ray Optics

Applications to Solids
Editor: H. J. Queisser
1977. 133 figures, 14 tables. XI, 227 pages
(Topics in Applied Physics, Volume 22)
ISBN 3-540-08462-2

Contents:
H. J. Queisser: Introduction: Structure and Structuring of Solids. – *M. Yoshimatsu, S. Kozaki:* High Brillance X-Ray Sources. – *E. Spiller, R. Feder:* X-Ray Lithography. – *U. Bonse, W. Graeff:* X-Ray and Neutron Interferometry. – *A. Authier:* Section Topography. – *W. Hartmann:* Live Topography.

Springer-Verlag
Berlin
Heidelberg
New York